PROJECT RAINFALL

PROJECT RAINFALL

THE SECRET HISTORY OF PINE GAP

TOM GILLING

ALLEN&UNWIN
SYDNEY · MELBOURNE · AUCKLAND · LONDON

First published in 2019

Copyright © Tom Gilling 2019

All rights reserved. No part of this book may be reproduced or transmitted in any form or by any means, electronic or mechanical, including photocopying, recording or by any information storage and retrieval system, without prior permission in writing from the publisher. The Australian *Copyright Act 1968* (the Act) allows a maximum of one chapter or 10 per cent of this book, whichever is the greater, to be photocopied by any educational institution for its educational purposes provided that the educational institution (or body that administers it) has given a remuneration notice to the Copyright Agency (Australia) under the Act.

Allen & Unwin
83 Alexander Street
Crows Nest NSW 2065
Australia
Phone: (61 2) 8425 0100
Email: info@allenandunwin.com
Web: www.allenandunwin.com

A catalogue record for this book is available from the National Library of Australia

ISBN 978 1 76052 843 0
Set in 11.5/17 pt Sabon by Midland Typesetters, Australia
Printed and bound in Australia by Griffin Press, part of Ovato

10 9 8 7 6 5 4 3 2 1

The paper in this book is FSC® certified. FSC® promotes environmentally responsible, socially beneficial and economically viable management of the world's forests.

In memory of
John Manchester

Contents

Introduction		ix
1	The space race	1
2	Decision No. 546	17
3	Pine Gapski	25
4	Memo to Mr Bailey	33
5	Cooks, bakers, computer operators	41
6	No need to know	49
7	Make it 'no comment'	57
8	The beauty of the domes	69
9	An Aussie bomb	75
10	Chinese whispers	89
11	Keep your hands off	99
12	The birds	105
13	Number two on the shit list	109
14	'I can't stand that [****]'	121
15	NSSM 204	137
16	Never again	149
17	Drop that hamburger!	155

18	Apocalypse now?	175
19	Reds, ratbags and radicals	193
20	No admission	203
21	A magnet for protest	213
22	Desert Storm	227
23	The canteen tour	237
24	Hunting for Osama	245
25	Snowden	257
26	Rainfall	267
27	A saucerful of secrets	273
Bibliography		301

Introduction

Type the words 'Pine Gap' into Google Earth and it takes you straight there, past the threatening road signs, over the perimeter fence and right up to the door of the top-secret operations building. The site is misnamed 'Pine Gap Military Facility Australia' but there is no mistaking the gleaming white radomes of the place officially known as the Joint Defence Facility Pine Gap.

In the 1960s enterprising journalists and cavalier university lecturers risked arrest and prosecution to get close enough to count the proliferating radomes (Gough Whitlam was allowed to visit in 1969 and miscounted them) but today you can zoom in close enough to do it on your laptop.

During the Cold War, when Pine Gap's main task was keeping an eye on Soviet missiles, peace protesters regularly made the trip to central Australia to demonstrate at the base, with activists leading police and security guards a merry dance behind the wire after breaking in with wirecutters. Since then, Pine Gap has become a vital cog in the American military machine, providing real-time battlefield information to commanders on the ground and locating 'targets' for assassination by US drones.

Once the subject of intense political debate, Pine Gap often goes unmentioned in the Australian parliament for weeks on end. The euphemisms used to obscure its secret purpose—vague references to its 'critical contribution to the security of both Australia and the US'—have scarcely changed in five decades.

No-one demurred when the defence minister, Christopher Pyne, declared in the House of Representatives on 20 February 2019 that Pine Gap 'represents one of the finest examples of collaboration, innovation and integration, and has delivered remarkable intelligence dividends to both our nations'. No-one pressed Pyne on the 'new demands and new challenges' facing Pine Gap or asked why it was still necessary that 'relatively few people' knew what happened there. On the contrary, the Labor opposition replied that it 'supports every word the minister has said'.

One person who knew what happened at Pine Gap was the late Professor Des Ball, who died in 2016, a few months before I began researching this book. Parliamentary committees frustrated by government stonewalling about Pine Gap and other US bases always knew who to call for accurate and up-to-date information.

Drawing on the technical analysis done by Ball and his colleagues at the ANU's Strategic and Defence Studies Centre, and on interviews and recently declassified documents from both Australian and US archives, *Project Rainfall* tells the story of Pine Gap from its genesis in the minds of America's Cold War intelligence chiefs to its modern role as a weapon in the Pentagon's 'war on terror'.

Insiders who have written about Pine Gap have had to submit to heavy censorship, while foreign whistleblowers such as Edward Snowden have been threatened with long jail sentences. More than five decades after the signing of the Pine Gap treaty, it remains true that the story of Australia's most secret place can only be written from the outside.

Chapter 1
The space race

The detonation of the first Soviet hydrogen bomb on 12 August 1953 sent shockwaves rippling far beyond the test site on the remote Kazakhstan steppe. The test, known as Joe-4 (it was the fourth Soviet nuclear test announced by the Americans), had an explosive power of roughly 400 kilotons of TNT—around 30 times the yield of the Hiroshima bomb. Although much less powerful than US thermonuclear tests conducted over the Marshall Islands, Joe-4 had—at least according to the Soviets—one big advantage: it was ready for immediate delivery by bomber.

The Americans understood at once the seriousness of the Soviet nuclear threat. The scientific advisory board of the National Security Agency entrusted a special study group to investigate. The top-secret Robertson report, compiled in 1953 and only approved for release in 2013, concluded that a surprise atomic attack on the United States 'would result in carnage, devastation, psychological shock, and curtailment of our retaliatory ability on a scale difficult to estimate or even to comprehend in terms of any previous experience'.

The same year the Continental Defense Committee, chaired by retired General Harold Bull, warned that existing US air defence plans were wholly inadequate and declared the need for strategic warning of a possible Soviet nuclear attack to be so great 'as to warrant any possible attack on the problem, regardless of its cost, funds and manpower'.

Existing radar and ground observer installations could at best give around 30 minutes' warning of an attack on coastal targets. Such a warning was only useful for a tactical response, since the attack would already be underway.

What the US government desperately wanted was to stretch the reliable warning period of a Soviet attack from minutes or hours to 'two or four days'. This, it was thought, would give the military time for 'the complete deployment of our offensive and defensive forces, which would increase the chance of turning the enemy's operation before its mission had been accomplished and might even induce the enemy to abandon the attack'.

The US air defence system as it existed in the early 1950s could not provide such a warning; in fact, previous studies had been undertaken on the understanding that, in the words of the Robertson report, 'no earlier warning will be available'. In order to obtain the desired early warning of Soviet attack, the US realised it would have to turn to 'other sources of intelligence'. As in the past, these could be divided into overt, covert and signals intelligence. While traditional overt and covert intelligence should be 'vigorously cultivated', the Robertson report warned that to rely on them would be to 'court disaster'.

Signals intelligence—'above all COMINT [communications intelligence]'—was judged to be the most promising source of strategic warning of an impending attack on the continental United States. Interception of 'high-level cryptographic systems' and their exploitation 'on a timely basis' therefore became the primary recommendation of the Robertson report.

The strategic importance of signals intelligence had been repeatedly shown during the Second World War, with the interception of high-level Japanese and German messages allowing the Allies to penetrate the enemy lines, enter government ministries and military headquarters, and gain vital early warning of impending operations. According to the CIA's 'History of SIGINT [signals intelligence] in the Central Intelligence Agency, 1947–70', Japanese and German military communications and secret diplomatic communications were 'an open book to the US Government', while the Battle of Britain, the invasion of Europe, the war in Africa, and the Pacific war were 'all fought to a large extent with prior knowledge of enemy capabilities and intentions attained through US-UK COMINT'.

General Marshall acknowledged that Allied code-breaking had been crucial to both Eisenhower's campaign in Europe and Macarthur's in the Pacific and by helping to shorten the war had saved countless American lives.

The cost and difficulty of building landlines in a country as vast as the USSR made the Soviets especially dependent on radio communications. This, it was hoped, would give CIA eavesdroppers the opportunity to 'penetrate the frontiers and to share the secrets of the Soviet government'.

The Soviet atomic arsenal had grown rapidly since the first Russian bomb was detonated in 1949. Four years later the Robertson report estimated that Moscow had an atomic stockpile (excluding thermonuclear bombs) of about 120 weapons, which was predicted to rise to 300 by 1955. The Soviet long-range bomber force was believed to have 1000 propeller-driven medium bombers, which the report judged to be 'capable of reaching all parts of the United States on a one-way mission'. These bombers, similar to the US B-29, were expected to be augmented by approximately 180 heavy bombers that would 'most likely be capable of reaching all targets within the US from Soviet bases on a two-way mission carrying atomic weapons'.

To counter this threat, the Americans had 53 fighter squadrons, each with 25 interceptors (only 15 per cent of which could fly in all weathers) and 57 anti-aircraft battalions. The interceptors were fitted with forward-firing guns. 'Our present air defence system has a kill probability of between 0 and 15%', the report noted grimly.

By expanding the defensive radar network across Alaska and Canada, the government hoped to be able to increase the warning time for a coastal attack from 30 minutes to two hours. More interceptor squadrons would be added and anti-aircraft battalions would be equipped with guided missiles instead of guns. The cost of upgrading was going to be enormous: billions of dollars a year.

By the end of 1960 an estimated US$40 billion would have been spent on buying 90 minutes of extra warning time and a kill probability of 'approximately 50%'. It was not nearly enough. Military planners were determined to have an effective strategic warning system, a system that gave them days, not hours. Cost, in the words of the 'History of SIGINT in the Central Intelligence Agency', was a 'secondary consideration'.

The CIA began experimenting with 'sky spying' in the mid-1950s. Early efforts involved floating balloons equipped with cameras over Russia. The development of the U-2 spy plane allowed the CIA to conduct day and night ultra-high-altitude reconnaissance, with the aircraft capable of loitering for long periods over Soviet territory while staying beyond the reach of MiG fighters. U-2 surveillance flights over the Soviet Union came to a sudden halt when a plane flown by Gary Powers was shot out of the sky by a Russian surface-to-air missile on 1 May 1960, just two weeks before a keenly anticipated east-west summit in Paris. (In a paper entitled 'American geosynchronous SIGINT satellites', published in 1993, Major A. Andronov, a Soviet military expert, claimed that American high-altitude spy planes were shot down not just over the USSR but over China, East Germany and Cuba. Andronov estimated that 'in the period 1950–1969 about 15 US and NATO reconnaissance aircraft were shot down'.)

The loss of Powers' U-2 embarrassed President Eisenhower and eventually forced the United States to admit its airborne spying program. Gary Powers survived, but the incident—and the attempted US cover-up—scuppered the Paris summit and accelerated the US–Soviet arms race. Eisenhower suspended flights over the USSR, a decision informed by the knowledge that a new piece of surveillance equipment was about to become available.

In August 1960 the US obtained its first photographs from a satellite over the USSR. The project was codenamed 'Keyhole' and the first satellite, KH-1, did everything it was supposed to. Although the first surveillance pictures were not as clear as those taken by the U-2s, the Keyhole camera photographed more Soviet territory than all U-2 missions combined. Its best resolution of about 40 feet enabled analysts to count individual bombers on Soviet airfields. At a stroke, the mission disproved the scaremongers' claim of a 'missile gap' that favoured the Soviet Union over the United States.

Information about the Keyhole satellites remained secret until the 1990s. In his article 'What we officially know: Fifteen years of satellite declassification', Jeffery Charlston writes that President Bill Clinton signed an executive order in 1995 instructing the CIA chief, Robert Gates, to release information about three 'space-based national intelligence reconnaissance systems' dating back to the early 1960s. They were known as the Corona, Argon and Lanyard missions. All three acquired photographic images from space and returned the film to Earth for processing and analysis.

Retrieval of the film canisters was fraught with difficulties. After being hurled into orbit aboard a Thor rocket, the satellite was manoeuvred into place by controllers at ground stations until its camera was pointing towards Earth. Exposed film was wound onto a spool in the re-entry vehicle, known as the 'bucket', for eventual return to Earth. When the reconnaissance mission was completed, the ground-based controllers would instruct the bucket to separate from the booster stage and use its own retro-rocket to position it for re-entry. After hurtling through the atmosphere, the bucket had to jettison its heat shield and guidance system and release a parachute before either being caught by a passing aircraft or scooped out of the ocean before it sank. Between August 1960 and January 1962 there were 26 Corona satellite flights but only 11 successful recoveries.

Camera resolution improved over the decade and retrievals became increasingly reliable as more hazardous land recoveries were abandoned in favour of mid-air 'catches'. As well as improving film capacity, engineers added a second bucket, allowing for images to be urgently returned to Earth midway through a mission. Weather, however, remained an intractable problem for optical photography from space, just as it did on Earth. Huge quantities of precious satellite film came back with nothing but pictures of Soviet clouds.

By the start of the 1970s, according to the 'History of SIGINT in the Central Intelligence Agency', satellite photography had become 'the principal US intelligence tool in denied areas', providing the US military with 'routine surveillance through regular detailed photography of known target installations'.

Australia had been allowed to see surveillance photographs from US satellites since the 1960s and was on the distribution list for material from the Keyhole satellites. In his book *Inside the Wilderness of Mirrors*, Paul Dibb writes that throughout the 1970s Australia received photography from the KH-9 Keyhole satellites, which was 'delivered . . . by US military aircraft to photographic interpreters in the top-secret seventh floor of JIO' (Joint Intelligence Organisation, superseded in 1990 by the Defence Intelligence Organisation).

Soviet airfields and missile sites were constantly monitored by orbiting Keyhole satellites. Steady enhancements in the quality of photographs enabled CIA intelligence analysts to zoom in not just on military structures but on vehicles and equipment. But the limitations of the technology—a finite film supply, the need to return the film physically to Earth for analysis, and the relatively poor resolution of the wide-angle camera—spurred engineers to hunt for new solutions.

The CIA and the US Air Force each ran their own photographic satellite programs. Since August 1960 the two programs had been coordinated by the National Reconnaissance Office (NRO). Based in the Pentagon, the NRO was reported to have the biggest budget of any US intelligence agency. By the end of the decade it was effectively in charge of managing and operating all the United States' secret satellite reconnaissance programs.

The performance of the Corona satellites steadily improved, but they all suffered from a basic limitation: the wide-angle cameras were designed to search large areas for objects of significance, not to investigate specific targets. Increasingly, the CIA relied on intercepted Soviet signals to identify targets for its spy satellites. These signals came from sources including radio communications, electromagnetic radiation, telemetry signals (such as those emitted during missile tests) and radar. Of these, the CIA came to consider telemetry the most important.

Among the crucial electronic data sent back to Earth during a missile test was information about rocket-motor thrust, fuel consumption and guidance systems. Data about critical events such as the firing of explosive bolts to separate different stages of the missile or to release the warhead (or warheads) would also be conveyed by telemetry signals. Listening in on this continuous flow of missile data enabled CIA telemetry analysts to draw conclusions about the number and size of warheads carried by a given type of Soviet missile, the missile's range and the likely accuracy of the warhead after separation.

Much of the signals traffic generated by Soviet missiles was intercepted by ground stations operated by the CIA in countries bordering the USSR. Some of the most important sites were in West Germany and Iran. Although information about the specific purpose of such sites was highly classified, the CIA was allowed to share some information obtained from these stations with the host country.

While the CIA was tasked with penetrating the Soviet strategic nuclear weapons program, overall control of signals intelligence lay with the National Security Agency (NSA), a shadowy organisation that was to play a central role in the development of Pine Gap.

Created in total secrecy in November 1952 on the orders of President Truman, the NSA is still considered to be the most secret of all US intelligence agencies. The text of the so-called Truman memorandum remained classified for decades. Much of what became known about the agency only came to light as a result of congressional scrutiny of its illegal activities.

Much larger than the CIA, the NSA is housed at Fort George C. Meade, a US Army base in Maryland. In 1980 it was estimated to have 130,000 employees spread across the globe and an annual operating budget of US$50 billion. In his book *A Suitable Piece of Real Estate: American installations in Australia*, Des Ball reported that the NSA operated 'more than $3 billion worth of decoding machines, scramblers, computers and virtually all known forms of electronic equipment ever built to gather and process information'.

Within five years of its formation, the NSA had to deal with a quantum leap in Soviet surveillance capability. The launch of Sputnik—the first man-made object to orbit the earth—in October 1957 heralded a wave of Russian satellite launches. Keeping up with the telemetry generated by these satellites threatened at times to overwhelm US intercept stations listening in to the signals traffic. Before long (according to the CIA history) the NSA was 'up to its ears' in magnetic tape filled with recordings of Soviet telemetry signals.

Five years after Sputnik, the US launched its first so-called 'ferret' satellite designed to complement the ground-based intercept stations by gathering electronic intelligence from space. The main purpose of the ferret system, according to Jeffery Charlston, was to 'collect radar emissions and identify radar sites'. In the event of nuclear war, accurate information about radar locations and operating features would enable American bombers to evade Soviet (and Chinese) air defences. The ferrets were placed in relatively low orbits, usually between 300 and 400 kilometres.

Such orbits were low enough to maximise the sensitivity of their signals monitoring equipment but high enough to ensure an orbital lifespan of several years. Effectively, the satellites could be relied on to function for as long as their batteries and recording equipment held out. Their usefulness was restricted, however, to intercepting electronic signals; the ferret satellites were no good, for example, for eavesdropping on Soviet telephone communications. For that, new satellite technology was needed.

In the summer of 1963 the *New York Herald Tribune* published an article about the Syncom satellite program, a collaboration between NASA, the US Department of Defense and Howard Hughes's Hughes Aircraft Company. The seeds of the program lay in an idea first put forward by the science and sci-fi writer Arthur C. Clarke, who imagined a system of communications in which signals were bounced from one ground control station to a satellite and then bounced back to a different station on the ground.

Unlike the Russian Sputnik and other low-orbiting satellites, which passed briefly over a given place on Earth before disappearing from view, the Syncom satellites were designed to occupy a much higher orbit—just below 36,000 kilometres. At such an altitude the relationship between the satellite and Earth fundamentally changes. Sitting directly over the equator, a satellite orbiting at 36,000 kilometres takes the same amount of time—24 hours—to orbit the Earth as the Earth takes to rotate once on its axis. To an observer on the ground, the satellite appears stationary: it seems to hover above a fixed point, never disappearing from view and never passing out of range of its ground control station.

A high orbit also has another advantage over a low orbit: from 36,000 kilometres roughly one-third of the planet is visible.

A US spy satellite stationed over the equator would be able to 'see' the whole of the central and eastern USSR.

The article in the *Herald Tribune* caught the attention of Albert 'Bud' Wheelon, a CIA officer who had joined the agency's Office of Scientific Intelligence in the late 1950s from TRW Inc., an aerospace company based in Redondo Beach, California. Reading the story, Wheelon realised that a synchronous satellite like Syncom could be used to intercept signals from targets inside the Soviet Union and relay them to a US ground station (although not necessarily a ground station *in* the US).

The sort of targets that quickly came to mind were telemetry signals from the Soviet missile test range at Tyuratam, in the Kazakh Soviet; the Plesetsk cosmodrome, a launch site for the R-7 intercontinental ballistic missile based south of Arkhangelsk and some 1000 kilometres north of Moscow; and the White Sea, a southern inlet of the Barents Sea used by the Soviets since 1955 to test submarine-launched ballistic missiles. Even Sary Shagan, another missile test site in what is now Kazakhstan, previously safe from US eavesdropping, would become vulnerable to interception by a geosynchronous satellite. Listening in on signals from these and other Soviet missile sites would give US intelligence analysts insights into technical details such as thrust and warhead configuration.

Also of interest to the CIA was the remote Kamchatka Peninsula. Photographs taken by a U-2 spy plane in 1957 had confirmed that the 1250-kilometre-long peninsula, which hangs like a half-eaten fruit from the northeastern tip of the Soviet Union, was not a missile launch site (as initially suspected) but a target zone for ballistic missiles fired from other parts of the USSR.

One of the problems with the low-orbit ferret satellites was that their 'fly past' was relatively brief; they were incapable of loitering above a target. As well as being able to intercept a wide range of emissions, the geosynchronous satellites Wheelon had in mind could be 'parked' above regions of special interest, enabling them to collect data continuously over long periods. But would they work?

Wheelon rounded up some of the CIA's top signals experts to determine whether or not Syncom could be the model for an entirely new system of spy satellites. One potential problem concerned the signals themselves. Transmitted at very high and ultra-high frequencies (VHF and UHF), telemetry signals did not bounce off the Earth's atmosphere but passed through it, leaking into space to be picked up by waiting satellites. But so did ordinary television signals. The risk was that missile telemetry signals might be drowned out by the noise of television chatter. Wheelon, who in 1963 became the CIA's first deputy director for science and technology, did not want to have the US government throwing money at a costly space-based surveillance system only to have his satellites beaming back the nightly TV news from Moscow.

Initial doubts about whether a geosynchronous satellite would be capable of performing the kind of mission Wheelon envisaged were laid to rest after he ordered a technical study. The results of the study were reviewed by a senior official from the Office of Special Activities, who summarised his findings in a memorandum on the subject of 'synchronous satellites'. Dated 4 October 1963, the now declassified memorandum concluded that it was 'technically feasible to develop a synchronous satellite capable of the missions described' and that the satellite would 'press the state of the art'. Wheelon was concerned, however, about the expense of the program, noting 'it is important that we proceed with the basic study as first step in a cost effectiveness review'.

It was not simply a question of how much Wheelon's satellites would cost but of who would pay for them. The synchronous satellite program was a CIA project but it would be funded by the National Reconnaissance Office. This arrangement was the cause of a long-running feud between Wheelon and the director of the NRO, Brockway McMillan, who did his best to undermine the program. In a memorandum dated 5 November 1964, McMillan claimed that it would be 'relatively easy' for the Soviets to disable the geosynchronous satellites simply by jamming the frequencies on which they operated. For this and other reasons McMillan concluded that 'the development of a synchronous satellite ... will probably not be justifiable', while reassuring Wheelon that the CIA's detailed technical studies would be viewed as 'an important source of information' for a competing satellite project by the US Air Force.

Eventually Wheelon got his way, outfoxing his rivals in Washington to secure funding for the CIA's geosynchronous spy satellites. From now on the CIA would play a central role in the development of America's space reconnaissance systems. While relying on the air force for the launch, tracking and recovery of its satellites, the CIA would have the final say in determining the frequency of missions, choosing surveillance targets and analysing results. It was no accident that for five decades almost every chief of facility at Pine Gap would be a CIA veteran.

Although he was unable to prevent the CIA from developing its geosynchronous satellites, McMillan remained hostile to the program during his two and a half years as NRO director. In his final memorandum to the Secretary of Defense, dated 30 September 1965, McMillan wrote, 'it is hard for me to believe that a rational analysis of the usefulness of telemetry data, in comparison, say, to the direct usefulness of . . . data to be gotten by other SIGINT activities, would justify so large an expense'. To the end, he resented the expenditure of NRO money on a CIA project he considered to be ill-conceived, extravagant and lacking in 'intellectual rigor'.

The geosynchronous satellite program, described in top-secret briefings as a 'multi-purpose covert electronic surveillance system', was codenamed Rhyolite, after a volcanic rock containing glittering fragments of quartz embedded in ordinary igneous rock—a metaphor for the extraction of valuable intelligence from a mass of worthless electronic 'noise'. The satellites were to be designed and built by Wheelon's old employer, TRW Inc., and operated by the CIA. Victor Marchetti, a former assistant to the deputy director of the CIA, described the Rhyolite satellites as sucking up not only telemetry signals but military, diplomatic and other communications 'like a vacuum cleaner'.

In *Pine Gap: Australia and the US geostationary signals intelligence program*, Des Ball quotes the following passage from a book called *The Technology of Espionage*:

> By 1967 a new kind of spy-in-the-sky was designed . . . it was tentatively classified 'IS', for integrated satellite . . . It could pick up, record and transmit emanations from all land communication systems including radio and microwave telephonic transmissions. It also rigged with other varieties of electronic surveillance equipment which were sensitive enough to do the work of the hundreds of secret listening posts the United States Intelligence community had in such places as Turkey, Iran and West Germany. It was capable of duplicating everything those listening posts could do.

'Picking up' data was one thing; bouncing it securely back to Earth for analysis was another. The value of data intercepted by US satellites could be drastically compromised, or even neutralised altogether, if the captured signals were themselves intercepted during their transmission to Earth. If, for example, Soviet missile testers knew which telemetry signals were vulnerable to interception, they could respond with countermeasures such as encryption, selective jamming or reducing signal strength. Knowing which signals were being listened to would also allow the Soviets to draw conclusions about US surveillance capabilities. Des Ball noted that it was 'very likely that at least some of the Soviet telemetry encryption and suppression practices are designed not just to deny information to US monitors but also to elicit information about the US geostationary SIGINT satellite program'.

While CIA engineers collaborated with TRW on the technical side of the Rhyolite satellite project, the agency began searching for a secure site in a friendly and politically stable country that could serve as a ground control station.

In order to determine which of their telemetry signals (and which components of those signals) were being intercepted by US geosynchronous satellites, the Soviets would have to intercept the downlink from the US satellites to Earth. Preventing this became the main priority for the Americans as they hunted for a suitable site for a ground control station.

Based on the signal frequency of US satellite transmissions and the size of the downlink antenna, planners calculated that a ground station would need to be surrounded by a safe area at least 160 kilometres in diameter to be immune from Soviet interception. This immediately ruled out islands such as Guam and Diego Garcia, both of which would be vulnerable to Soviet interception by ships lying just outside territorial waters and by Soviet aircraft patrolling the area. A station in a friendly but populous country such as the Philippines would always be susceptible to covert interception by Soviet agents on the ground. The only place that satisfied the criteria of being impenetrable to air, sea or land-based signals interception would be a remote site in the interior of a sparsely populated continental land mass located somewhere between the 50th meridian and the 180th meridian. Only one country offered such a site: Australia.

Chapter 2
Decision No. 546

In early 1965 the CIA station chief in Canberra, William Caldwell, broached the subject of an Australian ground control station with the then secretary of the Department of Defence, Sir Edwin Hicks. Further details were given to Hicks in May and the defence minister was briefed the following month. A team from the defence department's Defence Science Division was told to look for a suitable site.

Even as the CIA hunted for a location for its satellite station, U-2 spy planes continued to photograph military installations in Asia. A CIA report declassified in 2012 reveals details of U-2 Reconnaissance Mission C425C, flown on 31 July 1965. The report, still heavily redacted, states that Mission C425C took off from Taiwan and flew over 'denied territory' in North Korea for 'approximately one hour and fifty-six minutes'. During the flight the U-2 'provided photographic coverage of 33 COMOR [Committee On Overhead Reconnaissance] targets and photographed two Mig-21 aircraft in interception attempts . . . The weather was reported by the pilot to be variable III (broken clouds), or slightly worse than the predicted Category II (scattered clouds).'

Preliminary interpretation of the photography indicated that the flight over North Korea covered '13 airfields; 9 military installations; 3 naval facilities; 1 BW/CW [biological warfare/chemical warfare] center; 3 electronic targets; and 4 complexes'. Of particular interest to the CIA analysts were the location and types of radar deployed in China and North Korea.

The evaluation of Mission C425C was based on information and data supplied from four sites: the Joint Sobe Processing Center at Okinawa; the National Photographic Interpretation Center in Washington DC; the 67th Technical Reconnaissance Squadron at Yokota; and the US Pacific Command ELINT [electronic intelligence] Center at Fuchu. Three of the four sites were in Japan, but the results of the U-2 mission were not shown to the Japanese. They were shared only with the UK, Canada, Australia and New Zealand, the other four members of the 'Five Eyes' intelligence-sharing network that had been established in 1947.

While CIA analysts pored over the results of Mission C425C, the Defence Science Division was narrowing its search for a suitable ground station site to the area around Alice Springs, in the Northern Territory. Central Australia was a perfect location for eavesdropping on radio transmissions from Southeast Asia as well as from the central and eastern parts of the Soviet Union. A ground control station in the heart of the continent would also be immune from interception by Soviet spy ships. This was essential because, as the former NSA employee David Rosenberg revealed in his book *Pine Gap: the inside story of the NSA in Australia*, the satellite downlink would initially be unencrypted.

Eventually the searchers settled on a natural basin in the MacDonnell Ranges less than twenty kilometres from Alice

Springs. It was a former grazing lease called Pine Gap valley. In late 1965 Australian and US engineers started surveying the area. It didn't take long for them to realise they had found what they were looking for.

In *The Wizards of Langley: Inside the CIA's Directorate of Science and Technology*, Jeffrey Richelson recounts the day in 1966 when Robert Mathams, head of the Scientific Intelligence Group of Australia's Joint Intelligence Bureau, drove out to the valley with three CIA men, Bud Wheelon, Carl Duckett and Leslie Dirks, to toast the beginnings of the Pine Gap project, which now carried the unclassified name Rainfall.

Carl Duckett was a radar engineer and former disc jockey who had become an expert in rocket telemetry while working on US ballistic missiles in the years after the Second World War. Leslie Dirks was another significant figure in the CIA's Directorate of Science and Technology and would play a key role in the Rhyolite spy satellite program and the development of Pine Gap. Mathams and the three CIA men, Richelson writes, 'drove out into the Australian outback, about twelve miles from Alice Springs. They passed through some low hills, took seats on the ground, and opened a case of red wine.' Richelson does not record how many bottles the four men consumed that day, amid the heat and dust and flies of central Australia, but it cemented a collaboration that would far outlive the duration of the initial treaty.

In June 1966 the then US Secretary of State, Dean Rusk, visited Australia for a conference of SEATO, the Southeast Asia Treaty Organization. While in Canberra he addressed the Australian Cabinet and sealed the deal to build a ground control station at Pine Gap.

The location and natural features of the Pine Gap valley made it a 'quiet' environment, with little electronic activity.

The station itself would be further insulated from electrical interference by a buffer zone of roughly twenty square kilometres. Access from Alice Springs would be via a single 'all-weather access road', to be built by Australia at an estimated cost of $247,000. (All roads within the facility were to be provided by the Americans.) Soviet agents and inquisitive embassy officials would find it difficult to pass unnoticed in an outback town like Alice Springs. To further protect the site against human intrusion, pastoral leaseholders were encouraged to keep 'visitors' off their land. Even today, the only way for a curious visitor without security clearance to see Pine Gap is by air or by clambering up and over the rugged red ridges of the MacDonnell Ranges.

Preliminary work on Pine Gap started soon after Secretary of State Rusk flew home from the SEATO conference. With it began a program of official deception and obfuscation that has continued ever since.

The same month, June 1966, it was announced that an access road was to be built to the Mereenie water bores. Locals had been calling for such a road for years but the Northern Territory Administration had repeatedly refused. Now, seemingly out of the blue, work got underway on a sealed road. However, the new road did not end at the bores but kept going, right to what are now the gates of the Pine Gap station.

The fact that some kind of high-tech facility was being built in the middle of the Australian outback could hardly be concealed; nor could the involvement of the Americans.

Work on the site had barely begun when it ran into problems. The land chosen for the installation was part of Temple Bar station, owned by local farmer Jim Bullen. The secret 'Interdepartmental Committee Report on the proposed Joint United States/Australian Defence Space Research Facility' described the site as 'generally unimproved light grazing land' and put a 'tentative' valuation of the land at $1940. The committee 'hoped that this land can be acquired by agreement with the present lessees under the Lands Acquisition Act'.

The government wanted to buy 900 acres, more than half Bullen's grazing land. Bullen had agreed to sell but was said to have insisted the new installation be a 'wholly Australian' defence project. He also wanted a fair price. When the government eventually made its offer—reported to be 52 cents an acre—Bullen refused to accept it. Work on the site came to a halt. What happened next was recounted some years later by a local surveyor, Des Nelson, in a letter to the *Centralian Advocate*. 'Jim had been offered compensation for the resumed land but was not satisfied with the offer,' Nelson wrote. 'On March 22, 1967 I was among a party of six which conducted a survey of the area in question. Four of us were members of Primary Industries Branch, Alice Springs. One officer was an inspector from Lands Branch, Alice Springs. The sixth person was a land valuer from Darwin. As a result of the investigation, Jim Bullen received an increase in the compensation. My diary states that we "saw where a US tracking station is being set up".'

Des Nelson's published account contradicts a story that has become part of the folklore of Pine Gap: that Jim Bullen was offered a measly 52 cents an acre 'take it or leave it' and sold his land only because the alternative was to have it seized by the government for nothing. In any case, it was not until March that the dispute was resolved and work on Pine Gap resumed, only to be interrupted again by heavy rains and a strike by local unions. Building did not get properly underway until August 1967.

(It would be another eleven months before Bullen saw his money. The terms of the deal were recorded in Hansard on

23 April 1969: 'Eight hundred and eighty-seven acres under agricultural lease was acquired from Temple Bar Station owned by Mr J. Bullen on 25th July 1968 with compensation of $5,950 including costs. Three thousand, one hundred and sixty-two acres under pastoral lease were acquired from Owen Springs Station, then owned by Mr Milnes, Mrs Graham and the late Mrs Hayes, with a price paid of $1,035 including costs . . . The remaining 325 acres were unalienated Commonwealth land.')

As well as building a sealed road connecting Pine Gap with the main highway (and hence with Alice Springs and the airport), the department of defence was responsible for supplying the new installation with water. The secret 'Interdepartmental Committee Report' estimated that this would require 'the sinking of three investigation bores and two deep wells at a site about one mile from the . . . boundary at an estimated cost of $69,000'. The Americans were to provide 'all pumping equipment, associated buildings and pipelines to the site'.

As with the sealed road to the highway, Washington expected the Australians not only to do the drilling but to pay for it, although the Americans were said to be 'prepared if necessary to meet the costs themselves'.

As for housing the American employees, the US Government 'expressed a clear preference to avoid becoming involved in ownership and maintenance of substantial real estate in Alice Springs' and was 'prepared to pay the high rentals' needed to pay off the cost of new housing over the initial ten-year duration of the contract.

A Cabinet minute marked 'Top Secret' and dated 21 September 1966 records that the minister for defence 'informed the Cabinet of proposals for the establishment of a joint US/Australian Defence Space Research Facility at Alice Springs which a visiting American team was expected shortly to bring forward'. The Cabinet 'indicated in principle its willingness to support the proposal if it came forward'. Fifty copies were made of the Cabinet minute on what was officially designated 'Decision No. 546'.

Chapter 3
Pine Gapski

The United States was not the only nuclear-armed superpower interested in tracking satellites from a base in Australia. In 1960 the Soviet Academy of Sciences put forward a proposal for an Australian space tracking station. Two years later a CIA document entitled 'The Soviet space program' reported that the USSR was 'seeking to acquire sites for space tracking stations in Chile, Indonesia, Africa, and Australia'. Remarkably, the Americans appeared unconcerned by the prospect of a Soviet ground station in Australia tracking Soviet and US satellites in space, despite the fact that the two superpowers had just come perilously close to war.

On 15 October 1962 US intelligence analysts studying photographs taken by a U-2 spy plane discovered that the Soviets were building medium-range missile launch sites in Cuba. President Kennedy ordered a naval blockade to stop any more missiles from reaching the island and threatened the Soviets with military action to end what he described as a 'clandestine, reckless, and provocative threat to world peace'. A face-saving compromise enabled both sides to pull back from the brink, with the Russians removing their missiles from Cuba and the US agreeing to remove some obsolete missiles of its own from Turkey and pledging not to invade Cuba.

Just three weeks later, on 20 November, the US State Department delivered a memorandum on its international activities to one of President Kennedy's closest advisors, Timothy J. Reardon. In a short section referring to Australia the now declassified document, designated 'Special Report 101', said:

> We have informed our Ambassador in Canberra that we would have no objection to an arrangement whereby the Australian Government operates a passive tracking facility for the USSR capable of tracking US satellites, subject to certain safeguards. While we would prefer that the facility be limited to passive capabilities, we recognise that continued Australian refusal to grant firm Soviet requests for command capability for their own satellites would be awkward for Australia, and we would not press our objections provided the Australian Government could make arrangements which would preclude Soviet interference with US operations.
>
> We do feel strongly, however, in this latter event, that the Australian Government should have full knowledge of the telemetry rather than being satisfied with a 'black box' arrangement whereby uncoded raw telemetry is sent directly to the USSR for processing. Only in this way, we believe, can the Australian Government be sure of knowing the precise activities of the station.

The willingness of both Canberra and Washington to countenance a Soviet space tracking station—or any other Soviet technical activity—in Australia was more surprising given that Britain and the US had important military installations in Australia, and that Australia had long been a favoured target of Soviet spies.

A secret CIA study of Soviet espionage techniques placed Australia top of a list of countries in which local Communist Party members were involved in spying. These domestic spies were often run by KGB handlers masquerading as diplomatic staff. The CIA report noted that diplomatic immunity had 'prevented the arrest of hundreds of Soviet intelligence operatives who have claimed immunity when caught in compromising situations. Instead of receiving long prison sentences, they have merely been deported to the Soviet Union.' Diplomatic immunity, it said, had also enabled the Soviets to install offices within their embassies where 'sensitive records of espionage activity can be maintained and where discussions, planning, and cryptographic work for intelligence operations can be carried on securely'.

The report gave the names of Soviet agents who had been exposed between 1942 and 1959 as spies 'while functioning ostensibly as diplomatic or other official representatives abroad'. The list included an embarrassing number of Soviet spies caught operating in Australia.

To Australian eyes, the most notable name on the CIA's list of Soviet spies was that of Vladimir Mikhaylovich Petrov, 'third secretary and acting VOKS officer' at the Soviet Embassy in Canberra between 1951 and 1954. (VOKS was a Soviet cultural organisation.) Petrov defected in Sydney after he and his wife, Evdokia, were unmasked as Soviet spies. Before she could talk, Evdokia was kidnapped by Soviet Embassy officials and bundled onto a plane for Moscow, only to be rescued when the plane touched down at Darwin and allowed to defect with her husband. The Petrov case provoked a royal commission that identified a number of Soviet agents in Australia, but not the major 'spy ring' some had suspected.

The ramifications of the Petrov affair continued long after the royal commission, not least for Petrov himself, who was terrified that he, like Leon Trotsky, would be hunted down and killed by Soviet assassins.

Among other Soviet spies identified by the CIA were:

- Filipp Vasilyevich Kislytsyn, a second secretary at the Soviet Embassy in Canberra, who organised an 'illegal apparatus' in Australia and 'stud[ied] members of parliament and the diplomatic corps';
- Yevgeniy Vasilyevich Kovalenko, third secretary at the Soviet Embassy in Canberra, who slipped out of the country in the wake of the Petrov affair;
- Semen Ivanovich Makarov, who was promoted from clerk at the embassy to third secretary to first secretary but was identified by the CIA as first State Security Resident;
- Valentin Matveyevich Sadovnikov, second and later first secretary, who took over from Makarov as State Security Resident;
- Georgiy Ivanovich Kharkovetz, press attaché, assigned to 'develop agents among contacts in correspondent, government worker and diplomatic circles';
- Yanis Eduardovich Plaitkais, attaché, assigned to 'work among Russian emigres';
- Aleksey Vladimirovich Vysselsky, press attaché and later third secretary, a State Security officer who returned to the USSR in November 1950;

- Viktor Nikolayevich Antonov, a correspondent for TASS (the Soviet news agency), assigned to target 'newspapermen, members of parliament'; and
- Nikolay Grigoryevich Kovaliev, commercial attaché, assigned to 'develop contacts in political and industrial circles'.

Despite Petrov's revelations about Soviet espionage in Australia, talks continued throughout the early 1960s about the possibility of building a Soviet space tracking station in Australia. Meanwhile, new weapons were being developed at the Weapons Research Establishment at Salisbury, north of Adelaide, and tested on the Woomera rocket range in South Australia as part of the local contribution to the top-secret Anglo-Australian Joint Project that had been established in 1946.

In 1963 the Russian plan for a tracking station was scrapped in the wake of another Soviet espionage scandal in Australia. This time the scandal involved Ivan Skripov, a KGB officer suspected of being at the heart of a spy ring operating in Australia. The most likely target of Soviet spies was the secret missile program at the Weapons Research Establishment. Among the missiles built at Salisbury and tested at Woomera was the British Blue Steel guided nuclear missile. Australian naval intelligence had detected Soviet submarines operating in South Australian waters at times that coincided with rocket launches at Woomera.

In February 1963 Skripov was declared persona non grata by the Australian government and ordered to leave the country within seven days, but other Soviet spies escaped discovery. Nevertheless, the ASIO director-general, Sir Charles Spry, was personally congratulated by the prime minister for his 'outstanding work'.

The intense media coverage that accompanied Skripov's expulsion seemed to have put an end to Soviet hopes for a space tracking station in Australia. But in 1965 Australian interest in the station was revived as part of a broader plan by the Menzies government to improve relations with the USSR. The idea went nowhere and in May 1966 the North American Air Defense Command's 'Weekly Intelligence Review' quoted the Australian ambassador to the United States, Keith Waller, saying that Australia had 'suspended action on a proposal to establish a Soviet space tracking station in Australia'.

Waller's statement appeared to finally kill off the proposal. The 'Weekly Intelligence Review' reported that 'the Soviets have not been able to establish a truly global space tracking network because of difficulties in arranging for the establishment of tracking stations in Free World countries'. A year later, however, *Newsweek* magazine reported that Australia 'might agree' to operate the first joint tracking station with the USSR in Western Australia 'if the US approved'. Without such a station, the Soviets could only communicate with their satellites via ships in the Pacific.

The US did not approve, and no Soviet space tracking station was built in Western Australia—or anywhere else. The Americans were worried about the Australian government being hoodwinked by Moscow into an agreement that prevented it from knowing 'the precise activities of the station' (an arrangement that was already working nicely for the Americans at Pine Gap).

But the Soviet dream did not die. In 1974 Gough Whitlam was asked in parliament whether the government had received a request to 'establish a joint Australian-Russian scientific base in Australia'. Whitlam acknowledged that 'a few weeks ago a party of visiting Russian scientists did raise with their Australian counterparts the proposal to establish a joint Australian-Soviet station'. He emphasised, however, that the proposal was for a 'station' not a 'base' and that it would be used 'for the purposes of photographing space objects and contributing to a study of the characteristics of the atmosphere'.

The US ambassador, Marshall Green, immediately protested. Whitlam assured the parliament that if there was 'military significance' in the planned Soviet station it was 'most unlikely that the proposals will be accepted'. US newspaper reports suggested that if Australia allowed a Soviet installation to be built, Washington might consider pulling out of ANZUS. The proposal went no further.

Only one more option remained. At the end of 1974 NASA was due to vacate its tracking station at Carnarvon on the Western Australian coast, 400 kilometres south of the US Navy's fleet communications centre at North West Cape, outside Exmouth. Built in 1963, the Carnarvon Tracking Station had evolved to become the largest tracking facility outside the continental United States and had played a key support role in NASA missions including Gemini, Apollo and Skylab. Its radar system, said to be the most accurate in the world, tracked manned space flights and deep space missions as well as scientific, defence and communication satellites. In April 1974 the government was asked by a Carnarvon shire councillor if the space station could be handed over to the Soviets after the Americans had gone. Writing 'on behalf of the employees', the councillor suggested to both Mr Whitlam and the premier of Western Australia, Sir Charles Court, that there was 'no reason why the station and its staff could not be offered to the Soviet Government' if the Americans had no use for it.

More worldly observers than the Carnarvon shire councillor were quick to note that a receiver as powerful as the one at Carnarvon could probably be turned to another use—such as eavesdropping on US satellites transmitting data to Pine Gap. In any case, under the terms of the agreement that covered operations at Carnarvon, the US retained all rights to the property and equipment until the agreement expired in 1980. The NASA tracking station ceased operations in December 1974, but there would be no Soviet flag over Carnarvon.

While Moscow never got its own Pine Gap, the Orroral Valley tracking station, built in the ACT as part of NASA's Spacecraft Tracking and Data Acquisition Network, would play a role in the 1975 joint US-Soviet project that saw Apollo and Soyuz capsules linking up in space. By then it was American spies, not Soviet spies, making trouble in Australia.

Chapter 4
Memo to Mr Bailey

During the weeks leading up to the official signing of the Pine Gap treaty, little thought was given to the question of how much information should be provided to the Australian public. On 6 December 1966, just three days before the public agreement was to be signed by the minister for external affairs, Paul Hasluck, an assistant secretary in the prime minister's department, A.T. Griffith, wrote a memorandum to his boss, P.H. Bailey. 'The presentation of the existence of this facility publicly poses some serious problems,' Griffith wrote, 'none of which in my mind seems to be satisfactorily resolved.'

The difficulty, as Griffith saw it, was that the bureaucrats charged with selling the benefits of Pine Gap had not been told what it was. 'Admittedly,' he told Bailey, 'it is difficult to do a good job on this without knowing the story but we must accept the decisions on this aspect and do the best we can. The American company primarily involved is well known for its activities in satellite communication. This, in my view, gives the key to public presentation.'

Griffith was worried by the description of Pine Gap as a research facility if (as he suspected) it also had an operational function:

> I do not know what is in Defence mind but as the plot is moving at the moment, on official paper at any rate, it is to be described as a defence space research facility. This means it is a non-operational facility. This is alright if it holds. The Americans have a tremendous propensity to talk about these things in some ways and there will be all sorts of questions put to the Government about the facility. The Government's answer is that it is a research facility. How long will such an answer appear fruitful? A flat question—is it a research facility or an operational facility—is all that is needed to make things really awkward.

In paragraph 6 of his memorandum Griffith outlined the dilemma the Australian government was facing: to come clean about the purpose of Pine Gap and risk compromising its military value or to deceive parliament and the Australian people and risk being caught in a lie. The 'essential problem' identified by Griffith was 'to present a story which is reasonably approximate and enables the Government to hold a sensible position without being basically untruthful before Parliament'.

> If we have to insist that it is research, we will not deceive the foreign governments so much as our own people. It is quite easy for the Russians to know what satellites are in the air and guess the nature of them. Hence, the Soviet will not be backward in feeding stories in. If we have a weak jumping-off position publicly, I do think we will have a difficult time.

We should remember that there are other kinds of defence activities in space which are not passive, but hostile, and that as the space war develops all sorts of ugly propositions will emerge about what is in space—thermo-nuclear bombs which can be directed to targets on earth etc. . . . we need to look at this matter very hard. The better the explanation, the easier it will be to accept parliamentary processing. Not to process the matter effectively in Parliament would be to make it fundamentally suspect.

A spin doctor before the term was invented, Griffith foresaw that Pine Gap would set the Australian government against the Australian parliament. He counselled against the fiction that the business of the Joint Defence Space Research Facility was research, not because it was a lie, but because it was a bad lie.

Griffith's solution was for the government to state that Pine Gap was involved in 'defence communications'. This had the virtue of being true (or at least half true) since the main US contractor, the Texas-based Collins Radio Company, was well known as a 'communication expert in space activity'. According to a special report entitled 'The corporatisation of Pine Gap', published by the Nautilus Institute for Security and Sustainability, Collins Radio was founded in 1930 to produce radio equipment for amateur enthusiasts and by the end of the war was producing a wide range of high-performance communications equipment. In the 1950s Collins was working with the US Navy and by the early 1960s it had moved into space, successfully transmitting the first photograph via satellite (a picture of President Dwight Eisenhower) and delivering communications systems used in Project Mercury, the first US manned space flight mission. At the end of the decade the famous footage of

Neil Armstrong and Buzz Aldrin walking on the moon was relayed to Earth on Collins Radio equipment.

The pretence that the function of Pine Gap was 'defence communications' saved the government from an outright lie because, as Griffith conceded, 'communications admits of a defence role and an operations role'.

(In fact, as an Australian whistleblower, Leonce Kealy, would later disclose, Collins Radio did not live up to either its initial billing of prime contractor or its reputation as an expert in space communications. Despite opening an office in Alice Springs, Collins's role was limited to supervising early construction work and managing support services such as air conditioning. The company's Australian subsidiary was excluded from the most secret parts of the base. 'COLLINS RADIO personnel all work on the OUTSIDE perimeter of the Base, with only a handful of exceptions,' Kealy revealed. 'They are not cleared to Top Secret classification.')

Griffith's two-page memorandum, stamped 'Secret' and addressed simply to 'MR BAILEY', had an immediate impact. Rising through the ranks of the Canberra public service, the assistant secretary's concerns were passed to the first assistant secretary and from him to the secretary.

'The attached note from Mr Griffith raises some very pertinent questions about the next steps in handling the proposed defence communications establishment near Alice Springs,' Bailey wrote to his superior. (Between colleagues, there was no need to bother with the fiction of 'research'.) 'Quite soon,' he went on, 'important elements of the scheme will inevitably become public.' Like his subordinate, Bailey was worried about the prospect of government ministers deceiving parliament and being unable to defend themselves. 'In the long run,' Bailey advised, 'the greatest possible frankness in the early stages is likely to prove the best policy.' Although anxious to help, Bailey was forced to admit that 'Mr Griffith and I . . . are unable to be of great assistance in advising because of the secrecy which has surrounded the arrangements'.

Bailey was more sheepish than his political masters about deceiving the public and withholding information, urging the secretary to 'give careful thought, and soon, to . . . what is to be said about the nature of the project (I would prefer something as accurate as possible, because one can then defend a main issue, rather than be defensive about a "cover" story)—who, in fact, wants the secrecy, once the initial arrangements have been made?'

The departmental correspondence suggests there was still time for bureaucrats to devise an effective strategy to inform parliament and the public about the project at Pine Gap, but that time was running out. The government, however, had no intention of discussing Pine Gap with 'the greatest possible frankness'. Just three days after Griffith's memorandum to Bailey, a government cable was sent to the Australian Ambassador in Washington, Keith Waller. Marked 'FOR WALLER' and headlined 'SECRET', the cable said:

> THE AGREEMENT RELATING TO THE ESTABLISHMENT OF A JOINT DEFENCE SPACE RESEARCH FACILITY WILL BE SIGNED BY THE MINISTER IN CANBERRA TODAY, 9TH DECEMBER, 1966.
> 2. A PRESS RELEASE IS LIKELY TO BE MADE OVER THE WEEKEND AND THE TEXT, AS SOON AS IT IS AGREED, WILL BE FORWARDED TO YOU.
> 3. THE IMPLEMENTING ARRANGEMENT WILL BE SIGNED BY SIR EDWIN HICKS FOR DEPARTMENT OF

DEFENCE AND SENT TO ARPA [Advanced Research Projects Agency] FOR SIGNATURE.
4. COPIES OF THE AGREEMENT (UNCLASSIFIED) AND THE IMPLEMENTING ARRANGEMENT (SECRET) WILL BE FORWARDED TO YOU.

The public agreement was concocted and announced as a front for secret treaties that would determine how the facility was to be operated, and by whom. These secret treaties were largely the work of two senior CIA officers, Victor Marchetti and Richard Lee Stallings. Stallings, who worked for the CIA's Office of ELINT (electronic intelligence), arrived in Australia in September 1966 and attended meetings in Canberra as the designated 'Project Chief'.

Signed in Canberra by Australia's minister for external affairs, Paul Hasluck, and the counsellor at the US Embassy, Edwin Cronk, the public agreement was to last for ten years, with the proviso that it could be terminated by either party with twelve months' notice. It referred to 'the establishment of a joint defence space research facility' and expressed the two countries' mutual desire 'to co-operate further in effective defence and for the preservation of peace and security'.

Articles 2 and 3 of the agreement stated:

> The Australian Government shall at its own expense provide such land in the vicinity of Alice Springs, Northern Territory, as is required for the purposes of the facility. All land so provided will remain vested in the Australian Government, which shall for the duration of this Agreement make the land available for the facility on terms and conditions to be agreed between the two Governments and shall for

this purpose accord to the United States Government all necessary rights of access to, and joint use and occupation of, the land.

The facility shall be established, maintained and operated by the cooperating agencies of the two Governments, and information derived from the research programs conducted at the facility shall be shared by the two Governments. These agencies are the Australian Department of Defence and the Advanced Research Projects Agency of the United States Department of Defense.

The public agreement is notable for its emphasis on 'joint' use of the Pine Gap installation; on 'cooperating' agencies; and on information derived from so-called 'research' being 'shared' by the two governments. Tellingly, the word 'joint' does not appear anywhere in the text of the secret 'implementing arrangement' (see Appendix 1).

The story that Pine Gap would be run by the Defense Advanced Research Projects Agency was a fiction designed to conceal its true purpose; Pine Gap was a CIA project and the CIA would decide how it operated. As Sharon Weinberger writes in *The Imagineers of War: The untold story of DARPA, the Pentagon agency that changed the world*:

> ARPA had almost nothing to do with Pine Gap's eventual operation, short of the occasional official's visiting to give the ARPA imprimatur . . . 'When I visited a foreign country not to be named, I was there as an announced "company man,"' recalled [Dr Stephen] Lukasik, the former ARPA director, who even forty years later declined to specify it was Pine Gap, or in Australia. 'I was the cover for their

station, the owner of what the sign at the gate called ARPA joint XXX Space Defense facility.'

On 11 December 1966, two days after the signing of a ten-year agreement with the US government, the defence minister, Allen Fairhall, announced that the sole purpose of the Joint Defence Space Research Facility would be 'research', the results of which would be 'available to both countries'.

The facility will be constructed on a site some twelves miles south-west of Alice Springs. About ten square miles of land will be required as a buffer zone to reduce electrical interference, although the facility itself will be built on an area of approximately fifty acres. The facility will include its own power plant, air-conditioned laboratories to house electronic equipment, and two radomes, each of which will enclose a large antenna. No launching or firing operations will be conducted at the site and it may be possible to continue grazing stock in the buffer zone.

Four months later, on 22 March 1967, at a meeting with Alice Springs residents, Richard Stallings said that the largest dish would be '100 feet in diameter'—a slight understatement (it turned out to be closer to 110 feet). Emphasising the 'scientific' nature of the facility, the CIA's top man in the territory also promised that there would be 'no serving military officers or men at the site'.

Chapter 5
Cooks, bakers, computer operators

Jim Bullen had worried that the Joint Defence Space Research Facility might turn out to be anything but 'joint'. Maybe he also suspected that the work it did would be anything but 'research'.

Two official committees were set up by the Australian government in Canberra to negotiate the terms by which it would operate. The first was a committee of permanent heads chaired by the secretary of the department of defence, Sir Edwin Hicks, and including the heads of Treasury, the attorney-general's department, and the department of external affairs. This was advised by a second, lower-level committee consisting of members of these three departments, along with others from the departments of territories; works; the prime minister; and the postmaster-general.

There were two priorities for the Australians: sharing the intelligence obtained at Pine Gap, and playing an active role in operations. The first appeared not to be a problem: the CIA professed to be willing to share both raw and processed intelligence with Australia.

In their 2016 paper, 'Australia's participation in the Pine Gap enterprise', authors Des Ball, Bill Robinson and Richard Tanter quote a document entitled 'Suggested requirements from an intelligence point of view' dated 20 December 1965. This stated that 'Information obtained... concerning Australia's area of main strategic interest... should be made available as soon as possible to appropriate Australian Intelligence Agencies', and that 'suitably qualified' Australian personnel would participate in the 'initial processing of data'.

In June 1966 the chairman of the lower-level advisory committee wrote a two-page minute outlining the progress of negotiations on 'the proposed new American project'. On the subject of 'Australian Participation', the minute noted that:

The activity may be divided into two parts:

(a) Hardware; and
(b) The processing of the results.

The Americans saw no objection in principle to Australian participation at the professional level in either (a) or (b). However in regard to (a), there would be difficulties in accepting Australian participation in the early stages. These difficulties were felt by the Australian party to be genuine.

With regard to (b) above, the Americans saw no objection to some Australian participation starting very early in the project.

Australian access to the results of the exercise was implicit in much that was said by the Americans and was also made explicit.

The second issue—the operational role of Australian staff—was not so easily resolved. Until Pine Gap actually began supplying intelligence, the Americans working there would be contractors rather than US government staff. As the contractors were responsible for the performance of the equipment after construction and installation, it was considered 'not practicable' for staff other than those engaged by the contractors to operate the equipment.

In early talks, the CIA could not—or would not—give any firm commitment on what operational role(s) Australians would play at Pine Gap. The Australian government accepted that operational roles would not be offered at once and was prepared to play a waiting game. The interdepartmental committee report accepted that 'Australian participation in the initial stages will for technical reasons be limited to logistic support personnel'. The committee expected, however, that 'Australia will, in due course, participate in the technical operation of the facility and obtain all the ensuing advantages of experience in these advanced techniques' and recommended that the United States 'should be asked to confirm that this expectation accords with American intentions'.

The CIA continued to stall. In September 1966 the CIA's 'Project Chief', Richard Stallings, gave what he described as his 'best estimate' of the 'numbers and types of Australian support personnel' likely to be needed from May 1968:

Services & Maintenance Manager 1
Clerks/Typists 2
Warehousemen 4
Vehicle Dispatcher 1
Drivers 8
Operating Engineer (Generators) 5

Maintenance Mechanics 6
Electricians 4
Plumber 1
Carpenter 1
Painter 1
Laborers 3
Vehicle Mechanic 1
Services Supervisor (mess & housing) 1
Cooks 4
Assistant Cooks 5
Mess Attendants 7
Stores Attendants 2
Recreation Stewards 5
Housekeepers 9
Janitors 5
Laundrymen 5
TOTAL 81

In addition, eleven more Australian staff would be needed from November 1968:

Cook 1
Assistant Cooks 2
Baker 1
Mess Attendants 3
Housekeepers 2
Drivers 2

In other words, Australians would be painting the walls, baking the bread and sweeping the floors, but operational work at Pine Gap would be done by Americans.

During negotiations the CIA had made little effort to hide its 'resistance' to Australian involvement in the operational systems at Pine Gap. The agency was happy enough to have a 'senior Australian official' (probably from the Department of Supply) on hand to supervise local staff and liaise with Australian authorities, but it did not want Australian hands near the equipment. While conceding that such resistance was 'understandable', the Australians agreed privately that it 'should not inhibit us from pressing for participation'. This meant focusing on the 'defence' rather than the 'space' aspect of the Joint Defence Space Research Facility. Instead of trying to get a 'space scientist' into the control room, government mandarins felt that the CIA might look more favourably on an Australian 'intelligence officer' with knowledge of signals intelligence (SIGINT). An Australian memorandum written in October 1966 outlines the likely appeal of such a person:

> The unique feature of the project is its information gathering potential; this can best be exploited by intelligence officers with appropriate electronic qualifications—people who know what kind of information to look for and how to interpret it... [SIGINT] is already a major subject for intelligence research in Australia and an appropriate Australian participant—one who is knowledgeable in this subject and has electronic qualifications—would not only be able to take part in the information gathering and interpretation function of the project but could also maintain liaison between the project and the Australian intelligence group studying this subject [i.e., the Defence Signals Directorate]... the purely satellite component of the project will remain essentially constant but... the information

component will undoubtedly expand in both scope and in technical complexity. This expansion will provide increased opportunity for Australian participation, and will ensure that we have full knowledge of all developments and get the best return for whatever investment we make.

Other documents confirm that the US government envisaged Pine Gap as an American facility, staffed by Americans. A document held at the National Archives, dated 17 October 1966 and marked 'Secret', begins:

The Embassy of the United States of America presents its compliments to the Department of External Affairs and has the honor to bring to the Department's attention a proposed program for general defense research in the space field . . .

While the construction of the facility would be covered by an unclassified agreement, the 'purpose and type of research and the resulting information would be classified'. On the subject of staffing, the document notes that the facility would be 'largely self-contained with United States personnel'.

In fact, the text of the 'implementing arrangement' made it clear that opportunities for Australian personnel at Pine Gap would be limited. Article 9 stated that ARPA 'agrees to provide appropriate United States personnel . . . to manage and administer all United States activity at the facility, and to carry out the technical operation of the facility'. Australian personnel, it said, 'may be required to perform other functions'.

Harold Holt won a landslide victory in the 1966 federal election under the slogan 'Keep Australia secure and prosperous—play it safe'. A month before the election the treasurer,

William McMahon, had complained to Holt about 'financial arrangements' relating to the Joint Defence Space Research Facility. It was not so much the money that annoyed him as the fact that discussions appeared to have been carried on behind Treasury's back.

The US government had initially guaranteed to pay all the costs of Pine Gap, with Australia's commitment limited to supplying the land. In the course of negotiations, the US position had shifted, partly as a result of Canberra's decision to levy income tax on Americans who stayed in the country for longer than eighteen months. Washington was still prepared, if necessary, to foot the whole bill. However, the counsellor at the US Embassy, Edwin Cronk, told Holt during a private meeting that the United States had changed its mind and 'would like to have an Australian contribution'. The amount Washington had in mind was $2.3 million in building costs plus $40,000 for the 'annual recurring costs' of the Commonwealth Police. When the deal was put to McMahon he objected, citing the 'lack of real evidence about potential value of the output of the project to Australia'.

Holt waited until the election was over before replying to McMahon's letter. The question was not how much Pine Gap was going to cost Australia, Holt explained, since Washington had made it clear that 'it would, in the last analysis, meet all costs associated with the project'. The question, rather, was how much Australia wanted to pay, and what it could expect to get in return.

> Having regard to the views which were expressed by our colleagues generally in the Foreign Affairs & Defence Committee, to the effect that it would be to our ultimate

advantage to pay certain costs . . . I had no real hesitation in adopting the proposition that Australia should meet certain costs, and thus concurring in what Mr Cronk referred to as a package deal.

Like many package deals, the Joint Defence Space Research Facility would end up costing a lot more than was advertised.

Chapter 6
No need to know

As the first signs of building work began to be noticed around Pine Gap, the Labor MP for Alice Springs, Charlie Orr, stood up in the Northern Territory Legislative Council and listed some concerns about the proposed 'military base'. He worried that the cost of living and property rentals would go up; that Alice Springs would become 'Americanised'; that there would be too many single men; and (paradoxically) that local schools would be unable to cope with the influx of children. Orr felt that the project was undemocratic and that the people of Alice Springs had been 'deceived' about what would be going on inside the facility.

The echo of Charlie Orr's ratbag rhetoric soon reverberated in Canberra. Seizing the opportunity to embarrass Harold Holt's Coalition government, Labor's Senator Reg Bishop demanded to know whether the people of Alice Springs had been conned into handing over their land for something that was not a 'space research project' but rather a base ('the biggest outside the United States of America') for the US military's 'spy in the sky' program. Were they right, he asked, to be 'concerned about its location in view of any future war?'

On 13 April 1967 Tasmanian Senator Norman Henty delivered the government's answer. Speaking for the prime minister, Henty insisted that there could be 'no grounds whatsoever for any assertion that the people of Alice Springs have been deceived as to the establishment or operation of the facility'. Henty also said that there was 'nothing about the operation or location of the station to support any suggestion that the installation would be singled out for special attention in time of war'.

Henty's statement was a blustering counterpoint to Fairhall's pastoral vision of sheep grazing happily in the buffer zone, but neither was convincing. The growing controversy over what was happening at Pine Gap prised the lid off a Pandora's box of military and semi-military collaborations with the United States that the Commonwealth government was now forced to disclose. On 29 August 1967 the Labor left-winger Dr Jim Cairns asked the defence minister a series of questions. They included:

> How many American military or other installations are in Australia, and where are they?
>
> Which are used for military and which for non-military scientific purposes?
>
> Does Australia exercise any control over these installations? If so, what control?

Six months earlier Cairns had narrowly missed becoming deputy leader of the Labor Party, losing by just a couple of votes to Lance Barnard on the same day that savage bushfires swept across southern Tasmania, killing more than 60 people and leaving thousands homeless. A fierce opponent of the

Vietnam War, to which the Menzies government had committed fighting troops in 1965, Cairns had no desire to see Australia becoming a military outpost of the United States.

The list of US installations already on Australian soil proved to be remarkably long. In his answer to Cairns's question, Fairhall listed the following:

> United States Navy Communications Station, North West Cape;
> Joint defence space research facility and a geophysical research project at Alice Springs;
> Deep space station at Island Lagoon, South Australia;
> Baker Nunn photographic satellite tracking station at Island Lagoon, South Australia;
> Tracking station at Carnarvon, Western Australia;
> Deep space station at Tidbinbilla, Australian Capital Territory;
> Satellite tracking and data acquisition network facility at Orroral Valley, Australian Capital Territory, and its associated mobile equipment station at Darwin, Northern Territory;
> Applications technology satellite station at Cooby Creek, Queensland;
> Tracking station at Honeysuckle Creek, Australian Capital Territory;
> Research station into atmospherical disturbances at Amberley, Queensland;
> A satellite tracking station at Smithfield, South Australia; and
> A balloon launching station at Mildura, Victoria, for upper atmosphere observations.

The North West Cape station, Fairhall told the parliament, was solely for the purpose of military communications. The rest were 'research facilities' conducting both civil and military research. As for who controlled them, Fairhall admitted that North West Cape was 'controlled by the United States Navy', while all the others were either under Australian control or under 'joint Australian-US control'—a term that would take several decades to properly decipher.

In the same month that Cairns forced the Commonwealth government to reveal the extent of US–Australian collaboration in military research, the South Australian Labor premier, Don Dunstan, welcomed the news that Adelaide had been 'selected' by Washington as the 'base city' for the installation being built at Pine Gap. As well as welcoming the economic benefits he thought would flow to South Australia from Pine Gap, Dunstan argued that 'the Joint Defence Space Research Facility will play a significant part in maintaining world peace'. Not everyone agreed. Some thought it brought the threat of war closer.

While the Commonwealth government obfuscated, Sydney's left-wing *Tribune* newspaper warned in August 1967 that there was 'no doubt that Australia has become part of the "spy in the sky" plan of the US Government'. Without offering any evidence, the paper suggested that Pine Gap might be co-opted into the American 'bucket' system of retrieving photographic film from orbiting spy satellites. The plan, it said, was for film to be 'dropped in capsules in a 10-mile buffer zone around Alice Springs instead of being "caught" over the Pacific'. In fact, the US was already phasing out land-based recoveries in favour of more reliable mid-air interceptions. Fortunately for the Indigenous people living in camps on the outskirts of Alice Springs, the idea of scattering top-secret film canisters around the Australian outback did not go any further.

Identifying Pine Gap as 'part of the US plans of aggression', *Tribune* concluded that the installation 'would be a prime target in the event of a major world conflict'. Pine Gap was no longer just a 'facility', it was a 'target'.

Around midday on Sunday, 17 December 1967, the prime minister of Australia, Harold Holt, waded into a wild surf at Cheviot Beach, near the Victorian town of Portsea, never to be seen again. It was Holt's government that, twelve months earlier, had signed the Pine Gap agreement with the United States. Soon after becoming prime minister, Holt had infuriated Australians opposed to the Vietnam War with his notorious unscripted promise that Australia would be 'all the way with LBJ'. Criticised for being too subservient to US military demands, Holt trusted Washington to guarantee Australia's security. As far as Holt was concerned, the ANZUS alliance with New Zealand and the US was the cornerstone of Australia's defensive strategy. The American bases, in turn, were the cornerstone of ANZUS. 'There is a price to pay for the alliance,' Holt's defence minister, Allen Fairhall, would later declare, 'and the price we pay takes the form of the facilities provided at Woomera, Pine Gap and North West Cape'.

Five days after Holt's disappearance, President Johnson attended his memorial service in Melbourne. A classified CIA memorandum entitled 'The security situation in Australia' concluded there was 'little risk to the President in connection with his trip to Australia for the funeral of Prime Minister Holt. The police and security agencies of Australia are well-trained and competent. They will take all possible precautions to prevent any situation which might endanger President Johnson.

Of course, a solitary act by some demented individual cannot be ruled out.'

While the agency was concerned about political demonstrations, the memorandum noted that most Australians were 'well disposed towards the US. There are groups that on occasion have made vociferous (but peaceful) anti-US demonstrations. The chances of even this sort of unpleasantness will be minimized, however, by the solemn circumstances of the President's visit.'

Archival footage of Holt's funeral at St Paul's Cathedral shows President Johnson looking close to tears. Johnson had lost both a personal friend and a political ally.

Holt's death robbed Pine Gap of a powerful supporter in Cabinet. On the ground, however, the new base was rapidly taking shape.

Vast amounts of American money were rolling into Alice Springs, along with American cars that were flown in on military transport planes. David Rosenberg, a US intelligence officer working inside the base, recalled that the Americans 'often sold them to the locals when they left, almost always at a profit, with Alice Springs becoming inundated with left-hand-drive vehicles'.

To satisfy the Americans' love of gambling, one local contractor, Terry Lillis, set up an illicit casino 'complete with a roulette wheel and a blackjack table'. The cards were dealt by a Canadian croupier whose day job was as a geologist searching for oil and gas near the Mereenie basin.

Despite being warned to 'keep a low profile and stay out of trouble . . . so as not to draw unnecessary attention to the Base', the cashed-up and free-spending Americans could hardly help making themselves conspicuous in a town like Alice Springs. Competition with the locals for female company caused the odd fight and racial tensions between Americans and Aboriginal people sometimes boiled over. When the preliminary construction work was finished, local contractors like Lillis abruptly found themselves laid off and replaced by others, probably for security reasons, since it would have been unwise to allow any individual to know too much about the internal layout.

The major building work at Pine Gap was finished in early 1968 and after a walk-through of the base the Joint Defence Space Research Facility was formally handed over to maintenance and operations staff on the last day of April 1968. The heavy technical equipment began to arrive soon afterwards on board US military transporters landing at Alice Springs airport.

The famous 'golf balls'—technically known as radomes—soon began to rise from the red dirt. Mounted on concrete, the perspex spheres were designed both to shield the antennae within from prying eyes and to protect them from weather and dust.

By November 1968 the third and fourth radomes were already under construction. More would follow.

Among other facilities at Pine Gap, Des Ball lists 'luxury accommodation' in 'motel-like modern units' for Australian and American personnel living on site, a dining hall, canteen, bar, 'hobby shop' and a storehouse said to hold food supplies for a week. Multiple diesel generators ensured the complex was self-sufficient in power. When choosing a suitable site the CIA had stipulated a need for 'a million gallons' of water a day, but it was 'happy to settle for less'. Bore water was piped and stored in 'two huge dams' inside the perimeter.

A seven-square-mile buffer zone, fenced and patrolled around the clock by guards and Commonwealth Police, was enough to keep out all but the most determined intruders, while aircraft were forbidden to overfly the site or to approach within four kilometres.

Not much could be done about the Northern Territory's infamous flies. In his book, David Rosenberg recalls stepping off the plane at Alice Springs and being met by a 'biblical "plague" of flies'. One of his fellow passengers 'was wearing a white shirt, but his back was black with flies . . . I glanced over my shoulder to discover that the flies had taken an immediate liking to me as well'. It did not take him long, he writes, to 'become expert in the legendary "Aussie salute"—habitually brushing flies away with a wave of the hand'.

Pine Gap had been casually described in the press as a 'multimillion dollar project', but the government kept Australians in the dark about the exact cost of the facility. In January 1968 the *Adelaide Advertiser* would quote a final cost of nearly $200 million. The following month the respected US magazine *Aerospace Technology* mentioned an 'anticipated cost' of US$225 million. In 1975 this figure in turn would be dwarfed by an estimate of a billion dollars quoted by the *National Times* newspaper as the likely replacement cost of Pine Gap—a figure that was reckoned to have doubled five years later when Des Ball asked senior CIA officers what it might cost the US to duplicate the facilities at Pine Gap in another country. In short, the strategic value of Pine Gap to US intelligence was incalculable, and successive US governments have been willing to spend whatever was needed to maintain it.

Chapter 7
Make it 'no comment'

The new prime minister, John Gorton, was more sceptical than his predecessor about the benefits to Australia of military collaboration with the United States. Gorton made no secret of his intention to make changes to the defence policies he had inherited from Holt and Menzies. A recently declassified CIA intelligence bulletin dated 20 August 1968 reveals Washington's concerns about the incoming Australian prime minister's military priorities:

> Prime Minister Gorton has shown a renewed determination to withdraw Australia's ground forces from Malaysia and Singapore within the next few years.
>
> A major review of Australia's defense policy is under way and should be completed in several months. In this context [US] Ambassador Crook reports that remarks made privately by the prime minister on 18 August indicated that by 1971 he would order all Australian forces except a few advisers and a small air contingent to pull out of the area.

This would parallel the UK military withdrawal. Gorton stated that he could see no sense in a continued Australian troop commitment because by itself the force was too small to be of importance in an emergency.

Gorton's views conflict with those of External Affairs Minister Hasluck and Defense Minister Fairhall, who reportedly are still strongly in favor of the Asian mainland 'forward defense' policy of former prime ministers Menzies and Holt. Ambassador Crook, however, believes that little can be done to change Gorton's mind. [Next sentence redacted.]

There was no talk, however, of the Gorton government backing away from Holt's commitment to Pine Gap. And with Allen Fairhall staying on as Gorton's minister for defence, parliament would continue to be kept in the dark.

On 2 August 1968 Sydney's *Sun* newspaper published an article on its front page under the headline 'The secret of Pine Gap'. The *Sun* reported that a 'massive $200m highly secret United States defence complex' covering an area of ten square miles at Pine Gap in the middle of Australia had reached the 'testing stage'. Three weeks later Senator Lawrence Wilkinson asked in parliament whether this 'highly secret United States defence complex' was the same thing previously described by the prime minister as a 'Joint United States-Australian Defence Space Research Facility'. The reply from the prime minister was that they were one and the same place, but that the *Sun*'s description was 'an exaggeration'.

The risks of depending on the newspapers for information became clear when Arthur Calwell—replaced as Labor leader the previous year by Gough Whitlam—demanded to know whether it was true that 'the United States Government intends to convert the meteorological site at Alice Springs, which it has been operating for some years, into a nuclear site at an estimated cost of $200m, or some such figure'.

In fact, Calwell was confusing the *Sun*'s account of Pine Gap with the 'geophysical research project' near Alice Springs, a wholly American-staffed station whose seismic measuring equipment could be used to monitor underground nuclear explosions in the region.

Fairhall put him straight, while noting that 'there seems to be a considerable amount of confusion about this matter' and taking the opportunity to chastise the press for the 'most fantastic reports' that were circulating about the purpose of the new 'joint Australian-American defence space research facility which is now being developed near Alice Springs'. In his brief answer to Calwell's question, Fairhall stressed four times that Pine Gap was to be a 'purely ... experimental station ... not an operational station' and suggested that the quoted cost of $200 million 'might well be a flight of fancy'.

If Calwell and others were confused about operations at Pine Gap, the government was happy to leave it that way. 'I cannot tell honourable members what it will do, I can only tell them what it will not do,' Fairhall told the parliament. 'It will not do any of the things purported of it in recent articles published in the Sydney Press.'

Several months later, in response to a question from Jim Cairns, Fairhall shifted his position. 'The work of the Joint Defence Space Research Facility,' he said, 'will contribute materially to the effective defence of both Australia and the United States.' To Cairns's next question—'Is the Government taking action at Pine Gap and elsewhere that would make Australia a certain nuclear target in the event of nuclear war?'—Fairhall replied, 'No', echoing the prime minister's earlier assurance to the parliament that there was 'nothing about the operation or location of the station to support any suggestion that the installation would be singled out for special attention in time of war'. Neither statement was true.

If the parliament was prepared to swallow the government's line, the press was less obliging. In December 1968 the *Canberra Times* reported a journal article by Robert Cooksey, a lecturer in international relations at the Australian National University, in which he suggested that Pine Gap could be used by the US military for the 'bombardment of China'. According to Cooksey, 'when originally conceived in the early 1960s, Pine Gap was to be concerned with an orbital bombardment system but this was overtaken by the Outer Space Treaty negotiated between the US and the Soviet Union and abandoned'. Cooksey went on to outline how the US could use its own 'orbital' ICBMs to attack China. At the time, Cooksey wrote, the only way for the US to carry out a nuclear attack on China was by sending ICBMs over the Soviet Union. While it was theoretically possible for US missiles to follow a different orbital trajectory over the South Pole, the inevitable loss of accuracy made it unfeasible 'unless there was a station similar to Pine Gap which could direct the missiles toward the end of their target run'. Naturally, this strategic possibility made Pine Gap a nuclear target for Chinese missiles. While Chinese guided missile systems, Cooksey argued, were 'not yet very sophisticated', a volley of four thermonuclear warheads would probably be enough to 'eliminate Pine Gap' and 'could cause considerable radioactive fallout over Queensland if a westerly wind was blowing'.

The US government was quick to respond to Cooksey's speculations, with an official in Washington pointing out that

US strategic missiles were 'self-guiding . . . and any "external" guidance system based overseas would prove less efficient and more costly' as well as being 'easier to jam'. The Australian government, meanwhile, stuck to its mantra that Pine Gap was 'an experimental project only'.

Cooksey's apocalyptic vision of Australian-based operators raining American missiles onto communist China was far-fetched, but the thesis underpinning it was correct: the government's claim that Pine Gap's purpose was 'purely experimental' was a lie.

Cooksey developed his argument in a later article for the Melbourne *Age*. Australia, he wrote, had become 'increasingly important to the United States in aerospace matters. With this has come involvement in American nuclear weapons systems. This does not mean US bombers with nuclear payloads or missiles with nuclear warheads are based in Australia. Rather, in the missile age, tracking and communications stations are as essential to the US as rockets and warheads or the command post in the White House. Outside the US, Australia is the largest centre for American aerospace operations. Because of its technical and logistic facilities, its political stability and external security, this country provides the most suitable piece of real estate in such operations in the Southern Hemisphere.'

Choosing his words carefully, Fairhall told the parliament in February 1969 that Pine Gap conducted 'pure research into aspects of space phenomena which might have a bearing on the defence of Australia and perhaps of the free world'. Beyond that, he said, 'If the Government were to disclose any more [information] . . . it would be clearly indicating a direction of investigation which could be of assistance to this country's potential enemies. For that reason the project is covered by security and no further information will be given.'

Two months later Fairhall changed his mind and decided that further information could be given after all. Australia, he told reporters, would derive 'considerable benefit from the facility, will gain from our participation as a partner in a scientific project using advanced technology, and in addition, of course we'll have full access to data available in the operation of the station and access to its services'. As to the potential risks associated with having such a valuable facility on Australian soil, Fairhall said he could not see 'any set of conditions short of a nuclear world war which would cause this to become a nuclear target and then no more than half a dozen others in Australia. So I see no problem.'

It was becoming increasingly clear that the Gorton government had no intention of telling Australians the truth about Pine Gap. There was, however, another and more disturbing possibility: that the government did not *know* what the Americans were up to—and did not want to know. Fairhall declared it enough for the Australian people to 'trust the good intentions of the United States' and assured the parliament that 'the benefits . . . are all our way'.

But what exactly were the benefits? If Fairhall's comments about the Nurrungar space tracking station were a clue to his knowledge of Pine Gap, then he appeared to have no idea. Asked at a press conference in May 1969 to explain the function of the Nurrungar tracking station on the edge of Island Lagoon, just south of Woomera, Fairhall replied: 'The functioning of the station will make a contribution to free world defence, but I wish you would not ask me how.' Fairhall's performance during the press conference was ridiculed by the opposition when a transcript was later read out in the Senate:

Question: Does the station have a defensive or offensive capability, or both?

Fairhall: No offensive capability.

Question: When you said that this had no offensive capability, did you mean in the same way as North West Cape?

Fairhall: No, I just meant no comment.

Question: I thought you said 'no' to the question whether this had an offensive capability?

Fairhall: Withdraw what I said before. Make it 'no comment'.

Question: Was this discussed by the Prime Minister when he was in Washington?

Fairhall: Yes.

Question: Was it discussed with the President?

Fairhall (turning to Sir Henry Bland, the secretary of his department): Yes. I think he mentioned that, didn't he, in his statement?

(Bland answers 'no'.)

Question: Is the tenure for 5 or 10 years?

Fairhall (again looking Bland for the answer): Do you remember what we had in mind?

(Bland answers 'no'.)

Fairhall: I do not think I would like to comment on that but there will be a definite term specified.

The spectacle of the minister for defence being unable to answer reporters' questions about a major new defence installation that, like Pine Gap, was expected to be targeted by the Soviet Union in the event of war suggested a fundamental lack of seriousness in the government's strategic agreements with the US. Trying to excuse his performance, Fairhall insisted that it had been lighthearted and good fun. 'Please do not take me seriously,' he told parliament later. 'The press conference was called in a hurry.' Des Ball was less forgiving, describing it as 'an incredible display of ignorance and confusion'.

Others shared his ignorance. In December 1978 John Gorton would admit, 'I don't even know what Pine Gap is all about. I didn't then [1969]. I could have asked but it didn't arise. I didn't ask about it.'

There is evidence, however, that the US government was willing to supply information to Australia but was rebuffed. The *New York Times* reported that the secrecy surrounding the launch of an early-warning satellite in November 1970 had been 'prompted by one or more nations such as Australia where American space tracking stations are maintained'.

In April 1973 Brian Toohey wrote in the *Australian Financial Review* that '[t]he feeling in some quarters in Washington is that no harm would come if the Australian public were told officially about these aspects of the bases'. According to Toohey, 'Pentagon officials claim that they have been more willing to assist with the political problems surrounding the bases than is generally recognised in Australia. They say that, soon after the [Whitlam] Labor Government came to power, the US offered to send a team to Australia to talk about the bases, but the Australian Defence Department refused the offer . . . They also claim that, under the previous Government, offers to provide greater access to the bases by Parliamentarians were turned down at the Australian end.'

The Gorton government certainly had no intention of throwing open the gates of Pine Gap to opposition MPs. Asked by Labor's Len Devine if he would declare 'an open day for members of this House and scientists interested to inspect the establishment', Fairhall replied, 'The simple answer is no.'

The government's prohibition on visits to Pine Gap further polarised an already divisive issue and entrenched a political antagonism that would have dramatic repercussions after Labor came to power in 1972. Not surprisingly, many MPs were insulted by the insinuation that they could not be trusted. As Gordon Bryant put it:

> I have no doubt the Minister for Defence will be able to visit Pine Gap. In what way has the Minister for Defence shown that he has any greater integrity, any more loyalty or any greater capacity to be a decent and dinkum Australian than any other member of this House? I am astonished that honourable members opposite should ... allow themselves to be treated in this way, with none of them capable of being trusted to visit Pine Gap, none of them to be trusted with security service information and none of them to be trusted at all, whether they sit on the left or the right ... we are challenging some of the fundamental principles upon which the Australian Government is based, ministerial responsibility ... [and] ... the rights of citizens to know absolutely what they are charged with and how security and other matters ... are administered in this country. Where does the money go?

It riled opposition MPs already banned from visiting Pine Gap to find that the subject was also off limits for parliamentary debate. 'We can no longer trust the expertise or judgment of ministers, particularly Mr Fairhall,' Labor leader Gough

Whitlam railed to the ABC's *This Day Tonight* program. 'We ought to be told the role and function of any defence installation in Australia, whether Australian or jointly run . . . there is no excuse for mystery. We don't want to get into the situation where defence is . . . a taboo subject. Too many mistakes have been made . . . through this excessive secrecy by the Australian Government.'

More galling still was the knowledge that MPs were being denied information about Australia that was readily available overseas. 'You can learn more about all these installations . . . in American publications, in the US Congress and congressional committees than you can in Australian publications or by asking questions in the Australian Parliament,' Whitlam told the ABC. 'In Canada there is much more public debate in their House of Commons and their magazines about joint installations and facilities than the Government will permit in Australia.'

Whitlam's deputy, Lance Barnard, put it more bluntly: 'Pine Gap has been the target of considerable Russian propaganda broadcasts,' he told the parliament. 'It seems that the Russians know more about what is going on at Pine Gap than do the people of this country.'

The government compounded its ban on Australian MPs visiting Pine Gap with a similar prohibition on MPs visiting the new satellite tracking station at Woomera, although US congressmen were entitled to visit both sites. In April 1969, two and a half years after the signing of the ten-year public agreement, the *Canberra Times* was still referring to 'the mysterious Pine Gap installation outside Alice Springs'.

After being refused permission to visit Pine Gap, the federal president of the Australian Labor Party, Senator Jim Keeffe, announced in June that he would write to President Nixon. 'I will now write to the American President seeking a permit to visit the base as it is apparent that the Australian Government has no authority over the facility, even though it is on Australian soil,' Keeffe told the press.

Fears of a sinister cover-up were exacerbated by newspaper reports of a 'bomb-proof' school being built at Pine Gap. In May 1969 the *Tribune* claimed $250,000 had been spent on the 'first stage' of a school that 'could be turned into a bomb-proof shelter. What makes this school building so curious is that two ends of the school are constructed from solid rock, the whole structure is heavy concrete and the other two sides of the school are to be fitted with nine feet long, six inch thick concrete louvres, each weighing about nine hundred weight.'

It was not only parliamentarians who were hampered in their work by the deepening cult of secrecy around Pine Gap. Under the headline 'Drinks top secret' the *Canberra Times* reported that 'the secrecy surrounding the $200 million base meant that licensing inspectors could not be admitted to approve plans for licensed premises. Nor could they make regular inspections required under the law.' In November 1969 a bill was introduced in the Northern Territory Legislative Council to exempt Pine Gap from provisions of the Northern Territory licensing laws.

The Americans had hoped that by using Adelaide as a 'base city' for Pine Gap they could reduce the need to build amenities and infrastructure in Alice Springs, but the idea had always seemed half-baked. A little more than a year after the South Australian premier, Don Dunstan, trumpeted Adelaide's role in the project, the minister for defence announced that it was all over. Both partners now felt that 'all American families associated with the facility should, as far as possible, be housed in Alice Springs'. Previously the plan had been for the majority of US families to live in Adelaide but studies had shown that it was 'more economical, and certainly more satisfactory from the family viewpoint, for them to live at Alice Springs'. Another 120 houses would now be built in Alice Springs on top of the 100 already built for the use of Australian employees. The cost was rising.

Chapter 8
The beauty of the domes

On 4 September 1969 the Labor leader, Gough Whitlam, was granted permission to visit Pine Gap, although the government took care not to allow him anywhere near the technical areas. Other Australians wanting to know what went on at the Joint Defence Space Research Facility had to rely on the work of journalists and academics, most of whose information was obtained from US sources.

Two months before Whitlam's visit, the Melbourne *Age* and the *Sydney Morning Herald* had published a series of provocative articles by Robert Cooksey and Des Ball that depicted Pine Gap not as an experimental research facility but as an arm of US global nuclear strategy. 'To a high degree of probability,' Cooksey wrote later, 'we concluded it was now functioning as an Earth station communicating with and processing data for satellites.' He and Ball identified three types of satellite:

- operational reconnaissance (especially ICBM alarm) satellites on north–south polar orbits;

- Program 949 Integrated Multi-Purpose Warning/Reconnaissance satellites, one of which was already in 'shake-down cruise' (i.e. was undergoing performance and calibration tests); and
- satellite-ABMs (anti-ballistic missiles) now in the development stage. Recent evidence suggested a breakthrough was imminent and that they would not be nuclear-armed, so deployment would not breach the Outer Space Treaty.

After learning of Whitlam's visit, Cooksey submitted a request 'to visit Pine Gap as an academic revising for publication a manuscript about the installation's functions and their strategic and political implications'. While Cooksey did not expect to learn much from an official guided tour, he hoped it might 'end my nightmares that the installation is merely a set of cardboard dummies'. As expected, his request was turned down, so Cooksey decided to make an unauthorised visit. He recounted his trip in another article for the *Age*:

> My first view was the usual one for a passenger approaching Alice Springs airport—a glimpse of the largest radome, containing the 110-foot antenna.
>
> At dusk on that afternoon I made what has become the obligatory drive along the new sealed road to the main gate: 'Commonwealth Police: Pine Gap Station: Joint Defence Space Research Facility.' But I did not bother with any symbolic knocking at the gate.
>
> Two days later I flew to Hermannsburg, seeing the installation clearly from the air, from the north on the way out and from the south returning to Alice Springs airport.

It was on the first leg of this trip that I noticed Mr Whitlam had been in error when he announced at his Darwin press conference there are now three radomes at Pine Gap (Government statements so far admit to only two). There is in fact a fourth radome, placed unmistakably between the original two.

That afternoon I drove as arranged to Temple Bar station, 900 acres of which was resumed as part of the approximately seven-square-mile prohibited area.

Through the Temple Bar gate I drove south-west as directed, choosing the way to the left whenever the track forked. After a time I realised I must be lost but drove on until I could find a spot to turn.

Soon I came to a fence, more solid than the usual Territory cattle fence, with a well-graded dirt road on the other side. Then looking up I saw through the trees the radomes of Pine Gap.

This must be the perimeter fence of the prohibited area. I walked along the fence to the middle of the valley.

Half a mile to a mile away I could see the low, white concrete buildings and the almost free-floating radomes, gleaming whiter than snow in the afternoon sun. No picture can convey the stillness of the valley and the beauty of the domes.

A guard drove up the dirt road in a four-wheel-drive vehicle. As I wandered back to my car he radioed for relief, came through a gate and guided me to Temple Bar homestead.

As a fact-finding mission, Cooksey's attempted infiltration of Pine Gap was something of a disappointment, although Cooksey clearly took some satisfaction from pointing out Whitlam's mistake. But facts were (and are) only part of the Pine Gap story. Thanks to the government's obsession with secrecy, the facility evolved in a twilight zone between information and speculation that made it irresistible to UFO fantasists. Even academics such as Cooksey were not immune. 'Apart from local rumour,' he told the *Age*'s readers, 'there is every reason to believe there is a nuclear power station at Pine Gap.'

> The first two radomes, and probably also the third, operate on UHF (ultra-high frequency), thus requiring a very considerable power supply.
>
> But there has been no increase in the output of the Alice Spring's [sic] power station beyond that appropriate to the town's growth. Nor has there been any increase in the supply of diesel fuel by Commonwealth Railways that cannot similarly be accounted for.
>
> Further, when the site was being surveyed in 1965 there was some drilling, which was once assumed to be for oil and is now known to have been for sub-artesian and artesian water. And a nuclear reactor requires a considerable regular supply of water. The two swimming pools of local legend are perhaps for water storage.
>
> There is nothing particularly dangerous or sinister about small nuclear power stations. The U.S. has several at its open-to-inspection research facilities in the Antarctic.
>
> What is significant about the nuclear power station, plus other features of the installation . . . is that Pine Gap can be utterly independent.
>
> In extended nuclear war with a series of strikes—if it is not itself taken out—Pine Gap can function even with the rest of Australia destroyed.

Mixing questionable scientific 'evidence' with 'local rumour' and even 'local legend', Cooksey showed that academics, too, could be seduced by the air of mystery surrounding Pine Gap.

After Gough Whitlam's visit, the defence minister decided to 'relax' conditions governing visits to Pine Gap to allow inspections by Commonwealth and Territory parliamentarians as well as 'certain officials'. The decision, he said, stemmed from a 'personal' visit he had made to Pine Gap, during which he had noticed 'an excellent relationship between the United States and Australian officers at the facility'.

In truth, little changed. Security, the minister warned, would remain 'no less stringent than I have previously indicated'. Key areas would remain accessible 'only to those specially authorised on a "need to know" basis'. No photographs would be allowed. Labor MPs would be permitted to visit Pine Gap but they would still be barred from sensitive areas and they would not be allowed to reveal what they had seen.

The 'relaxed' access conditions took the pressure off the government in parliament by transferring responsibility for security onto opposition MPs, ensuring (as Cooksey put it) that the debate about Pine Gap would continue to be 'conducted mainly in newspapers, on television and within universities—not in Parliament and by political parties'.

Both the *Age* and the *Sydney Morning Herald* published editorials criticising the government for its excessive secrecy and asking what Australia stood to gain from the 'joint' facilities. While acknowledging that the series of articles by Ball and Cooksey in July contained 'a good deal of speculation', the *Age* warned that secrecy 'breeds suspicion that the Government is sheltering in silence, not for legitimate reasons of security, but because it fears public debate might prove politically embarrassing ... How much control does Canberra have over what goes on behind the barbed wire at Pine Gap? ... If an enemy did "take out" Pine Gap, or Melbourne, would the US Government put its own cities at hazard by retaliating on our behalf?'

The *Sydney Morning Herald* was sceptical of Ball's and Cooksey's more alarmist theories (in particular, that 'the Russians' might 'take out' Pine Gap before starting a nuclear war with the United States), conceding that their articles were 'of course, conjecture'. Its main concern, however, was with the public's right to know. While the *Herald* believed that 'a majority [of Australians] would choose to run the risks involved for the sake of preserving the nuclear deterrent', it noted that 'they have been told nothing either of the risks or of the advantages'. This, the *Herald* argued, constituted a 'denial of democratic rights. If the Government maintains that military security makes such a debate impossible, then it must expect that some critics will exaggerate the risks without mentioning the advantages.'

A public servant clipped both editorials for the files, but the government was unperturbed.

Chapter 9
An Aussie bomb

Less than two years after sealing the deal to build Pine Gap, the US secretary of state, Dean Rusk, had gone back to Canberra for a meeting with the Australian prime minister. By then Harold Holt was gone, either drowned in a deadly surf or (if you believed the conspiracy theorists) snatched away by a Chinese submarine. Australian and American soldiers were still fighting and dying side by side in Vietnam.

Rusk arrived on 6 April 1968. His talks with John Gorton, the maverick former fighter pilot, were expected to focus on Vietnam. Gorton, however, had other things on his mind. The United States and Great Britain, along with the USSR and another 50 nuclear and non-nuclear states, were preparing to sign the Nuclear Non-Proliferation Treaty, pledging not to transfer to other countries weapons, technology, or materials that could enable the production of nuclear weapons. Australia, which had sheltered for more than two decades under the US nuclear umbrella, was resisting. Unlike the passionately pro-American Holt, Gorton did not trust the US to protect Australia in the event of a nuclear attack. Australian scientists had already explored the possibility of developing nuclear weapons and some senior bureaucrats were keen for Australia to have its own bomb.

Signing up to the non-proliferation treaty would put an end to Australia's nuclear ambitions. Acutely aware of the instability in Southeast Asia, Gorton fiercely opposed 'giving up the nuclear option for a period as long as twenty-five years when [Australia] cannot know how the situation will develop in the area'.

Australian dreams of nuclear self-sufficiency dated back to the end of the Second World War. Nearly two decades before the CIA proposed building a satellite tracking station in the Australian outback, Canberra agreed to host a 'joint' military project with another of its wartime allies that would prove equally controversial.

British scientists had played a crucial role in the Manhattan Project that led to the Hiroshima and Nagasaki bombs, although the money spent by Washington dwarfed the British investment. In his article 'The British mission', Dennis Fakley writes:

> The British were amazed by the progress made in America and staggered by the scale of the American effort: the estimate of the total project cost was already in excess of one thousand million dollars compared with the British expenditure in 1943 of only about half a million pounds. [Senior British scientist Professor James] Chadwick was in no doubt that the first duty of the British was to assist the Americans with their project and abandon all ideas of a wartime project in England. He concluded that this would best be achieved by sending British scientists to work in the United States.

The Anglo-American collaboration came to a sudden halt after the war when the US government acted to safeguard its nuclear monopoly by withholding research, raw materials and test facilities from its former partner. After plans fell through for Canada to host a postwar Commonwealth atomic program, Britain turned its attention to Australia.

As well as scientific and military cooperation, the two countries were keen to collaborate in the increasingly sophisticated area of intelligence. The establishment in 1947 of the 'Five Eyes' intelligence-sharing network had been formalised by an agreement that divided responsibility for SIGINT collection 'between the First Party (the United States) and the Second Parties (Australia, Britain, Canada, and New Zealand)'. Their cooperation was not confined to SIGINT but included human intelligence collection and counter-intelligence as well as other clandestine activities. From the 1960s onwards, ground stations in the five member countries would download and relay intelligence from satellites.

While partnership in the Five Eyes network did not immediately defuse imperial tensions between Britain and Australia over status, it consolidated Anglo-Australian military collaboration at an important time.

In February 1946 the British government was given permission to use Woomera in South Australia as a rocket-testing site. In addition, it was granted sites at Montebello, an island group off the Western Australian coast, and at Maralinga, near Woomera, for testing atomic weapons. In 'Rethinking the joint project: Australia's bid for nuclear weapons 1945–1960', Wayne Reynolds writes that Australia's Labor prime minister, Ben Chifley, accepted the Anglo-Australian joint project 'subject to Australia having full access to information and being able to manufacture modern weapons at "a future date in accordance with the need to disperse manufacture through the Empire"'.

It was not just real estate that Britain needed for its atomic program but scientific brains and manpower. A vital part of Australia's contribution to Britain's atomic project was to be the creation of a national university capable of advanced scientific research. With its political and administrative infrastructure, and its proximity to the planned centre of the nuclear power industry, the Snowy Mountains scheme, Canberra made the perfect site for the new Australian National University. According to Reynolds, the Snowy was an 'ideal location for the construction of a plutonium-producing . . . fast breeder reactor' which would require 'vast amounts of water . . . for cooling and for moderation'.

Convinced that there was no chance of the United States allowing access to its nuclear test sites, Britain threw all its energy into the atomic collaboration with Australia. US observers were not invited to witness the Montebello test in October 1952. After a successful series of atomic tests in central Australia, Maralinga became a permanent test site, eventually hosting at least twenty firings.

The building of an experimental reactor at Lucas Heights, in Sydney's southern suburbs, was supposed to be the first step in a domestic nuclear program that within a decade would see the development of full-scale nuclear power reactors. Mark Oliphant, an Australian physicist who had played an important part in Britain's wartime nuclear research, was anxious that the government not close the door on Australia producing its own weapons. This did not mean actually building a bomb, Oliphant said, since atomic power plants producing plutonium and U-235 could be converted to the manufacture of atomic weapons 'in a matter of hours'. The manufacture of a thermonuclear weapon would require more sophisticated plant and installations but these 'could be tackled by any industrialised nation'. Oliphant had no doubt that 'Australia could best be defended by nuclear weapons'.

John Gorton agreed, arguing in 1957 in a speech to the Senate that Australia should secure its own 'atomic or hydrogen defence. I realize that a potential attacker of this country might be deterred by the possession of hydrogen bombs by the United States of America or Great Britain, but I think that we should be trusting very much indeed to the help that those great countries could give if we put our faith solely in a deterrent held by them.'

In his paper 'Surprise down under: The secret history of Australia's nuclear ambitions' Jim Walsh describes how Australian defence chiefs lobbied during the 1950s for an Australian bomb. When it became clear that the prime minister, Robert Menzies, had reservations, they pursued the matter behind his back. Menzies agreed, however, to let Britain test its nuclear weapons in Australia—a decision, according to Jacques Hymans, taken 'almost single-handedly . . . without consulting his Cabinet and without requesting any quid pro quo, not even access to technical data necessary for the Australian government to assess the effects of the tests on humans and the environment'.

Behind Menzies' offer was an expectation that Britain could be relied on for a nuclear guarantee that obviated the need for an 'Aussie bomb'. But, as Hymans points out, Britain was already winding back its regional security commitment. If anyone was going to offer Australia a nuclear guarantee it would be the United States, not Britain.

In September 1951 Australia had signed the ANZUS Treaty with the United States and New Zealand. James Curran notes in *Unholy Fury: Whitlam and Nixon at war*, that this was 'the first time that the Australians had entered into a treaty arrangement to which Britain was not a signatory'. ANZUS evolved from postwar plans for a broader and more ambitious Pacific alliance including Indonesia, the Philippines and Japan; Menzies had ridiculed this as an attempt to 'erect a superstructure on a foundation of jelly'. Although modelled on NATO, the ANZUS Treaty did not impose anything like the same responsibilities for mutual protection. In a letter to General MacArthur, Washington's chief negotiator, John Foster Dulles, made it clear that the United States 'can discharge its [treaty] obligations... in any way and in any area that it sees fit'. In short, Curran writes, ANZUS 'obliged the US to do not very much at all'. Protected by a treaty that, in fact, promised very little protection, Australia had to find other ways to align its defence interests with those of the United States.

The successful detonation of a nuclear device by China in October 1964 sent shivers through Canberra. Chinese cooperation with the Soviets had ended in 1959 when Khrushchev refused to supply Beijing with a prototype bomb. The 1964 test device was highly sophisticated and had been built without Russian help. With a yield of around 22 kilotons, it was roughly as powerful as the Fat Man bomb dropped on Nagasaki. In his article 'Isotopes and identity: Australia and the nuclear weapons option, 1949–1999', Hymans writes that the Australian government expected China to have a nuclear force 'capable of threatening a good part of Southeast Asia' by the end of the decade, if not before.

After the Chinese nuclear test the Menzies government worked hard to ingratiate itself with Washington. This meant bowing to pressure to support the US in Vietnam. Australia had been derided in the US Senate for its government's failure to match its bellicose rhetoric with troops on the ground. In June 1964 Senator Wayne Morse, a strong opponent of the war, ridiculed Paul Hasluck's decision as minister for external affairs to double the number of Australian military advisers in Vietnam:

> Despite its well-distributed press releases and the statements of its Foreign Minister, whose statement is an insult to the intelligence of anyone who will look into the facts, Australia's contribution to the war in Vietnam continues to be almost negligible.
>
> When its Foreign Minister advises that Australia is 'doubling' its advisers in South Vietnam, one would be impressed if one did not know that the figure being doubled is 30. In fact, if one listened to the Secretary of State in the Committee on Foreign Relations yesterday, he would have been impressed with Australia's role, until the facts were brought out as to what that role is.
>
> Listen to the facts as to what the role of this alleged ally of ours—Australia—is. And of any country or state in southeast Asia, which has a vital interest in South Vietnam, one would think it would be Australia. Also New Zealand, another alleged ally [that] has walked out on us in respect to South Vietnam. Who does the Foreign Minister of Australia think he is fooling? Why, Australia is increasing its advisory force from 30 to 60 advisers and you can be sure they will keep them away from the combat zone.
>
> I have not heard yet of any casualty among these Australian advisers. Furthermore we are not going to hear of any. Australia is perfectly satisfied to endorse instead the American war effort and encourage us to continue it

unabated. It is all right with Australia to have American boys die in South Vietnam.

In fact, when doubling the number of advisors Hasluck had also agreed to their switching to a combat role. Hasluck's decision, which broke from the previous policy of offering political support for the war while not being part of the fighting, led inevitably to Menzies' decision in April 1965 to commit a full infantry battalion to Vietnam. An article in US News and World Report in December that year described Australia's contribution to the US war effort in Vietnam as '1,300 troops, all-out moral support' and New Zealand's as '300 troops, heavy moral backing'. The British contribution was 'No troops, much sympathy' while France had given 'No troops—or sympathy'.

Australia's military support, while negligible beside the 175,000 American soldiers already in Vietnam, represented an important act of solidarity with Washington. In January the following year Menzies retired as prime minister and was succeeded by his protégé Harold Holt, who more than quadrupled the Australian troop commitment. Like Menzies, Holt was staunchly anti-communist and pro-American. Both understood that Australia's security depended on being able to shelter under the US umbrella, and that guarantees of American protection came with a quid pro quo. Sending troops to Vietnam was one down-payment, but another was about to fall due. In 1966 Holt's Coalition government agreed to host 'joint defence facilities' with the Americans at North West Cape, Nurrungar and Pine Gap.

Within two years, however, the security situation had shifted. Americans were growing sick of Vietnam. President Lyndon Johnson's announcement, on 31 March 1968, of a partial halt to the bombing of North Vietnam, and of his decision not to seek re-election, signalled a wider American disengagement from the war. Less than a week later Johnson's secretary of state, Dean Rusk, flew into Canberra for talks with the new Australian prime minister, John Gorton.

The ferocity of Gorton's tirade against the proposed non-proliferation treaty took Rusk by surprise. In a secret cable to Washington Rusk caricatured Gorton as behaving 'like [French President Charles] de Gaulle in saying that Australia could not rely upon the United States for nuclear weapons under ANZUS in the event of nuclear blackmail or attack . . . I will not recount here what I said to him but I opened up all stops.'

Annoyed by what he described as 'all sorts of picayune problems', Rusk urged the State Department to send a special mission to Australia led by 'someone who is thorughly [sic] competent on the technical side on nuclear matters and of the non-proliferation treaty to eliminate as many as possible of the misunderstandings and irrelevancies which are floating around the Australian Government'. If that was not practical, Rusk said, then it would still be possible to 'invite a delegation from Australia to visit Washington but my guess is that they would simply reflect the existing bickering inside the Australian government and would delay matters even further'.

While Gorton offered tepid Australian support for the 'idea' of a non-proliferation treaty, Rusk foresaw (correctly) that Australia would prevaricate about signing the treaty and would 'for a substantial period, hide behind objections raised by other delegations'.

Gorton's reservations about the non-proliferation treaty were not merely political but technical: he feared that signing the treaty might cause Australia's atomic energy industry to be 'frozen in a primitive state'.

Gorton and the head of Australia's Atomic Energy Commission (AEC), Philip Baxter, were both keen to pursue development of an Australian bomb. According to Hymans, scientists at the AEC had worked with officials at the department of supply to draw up precise cost and time estimates for atomic and hydrogen bomb programs.

> They outlined two possible programs: (1) a power reactor program capable of producing enough weapons-grade plutonium for 30 fission weapons (A-bombs) per year; (2) a uranium enrichment program capable of producing enough Uranium-235 for the initiators of at least 10 thermonuclear weapons (H bombs) per year. The first plan . . . was costed at an affordable $144 million (in Australian dollars) and was thought to be feasible in no more (and likely fewer) than seven to 10 years. The second plan weighed in at $184 million over a similar period.

Aware of opposition inside and outside the government to the idea of an 'Aussie bomb', Gorton played down the military aspect and argued instead for the economic benefits of a nuclear power program.

Rusk and the US delegation quickly smelt a rat. Exasperated by a series of objections they believed to have been cleared up during NPT negotiations in Geneva, the Americans tried to allay Canberra's fears of being left behind in the development of nuclear power. Among other things, they argued that the 'so-called peaceful "spin-off"' from the development of nuclear weapons had been obtained 'at great cost long ago' and was available to countries like Australia in open technical literature. In short, the Americans found it 'difficult to believe Australia would choose to spend hundreds millions dollars to "rediscover" what is now generally known'.

In his Washington cable Rusk explained that the object of the special emissary would be 'to deal with as many of the technical problems as possible in order to clear away that underbrush rather than to take on the major political issue as to whether Australia should give up the nuclear option'.

Rusk's advice was followed and a US mission did visit Canberra. At the end of April 1968, US arms control officials briefed a top British diplomat in Washington about the Canberra mission. Their view of the Australians was much more positive than Rusk's. Officials from Australia's Atomic Energy Commission impressed the visitors with 'the confidence of their ability to manufacture a nuclear weapon and desire to be in a position to do so on very short notice'.

> These officials seemed to have studied the draft NPT most thoroughly . . . The political rationalization of these officials was that Australia needed to be in a position to manufacture nuclear weapons rapidly if India and Japan were to go nuclear . . . the Australian officials indicated they could not even contemplate signing the NPT if it were not for an interpretation which would enable the deployment of nuclear weapons belonging to an ally on Australian soil.

Eighteen months after Rusk's fractious visit to Canberra, Gorton called a general election. He openly declared his commitment to a nuclear-powered (if not a nuclear-armed) Australia, announcing at his campaign launch that 'the time for this nation to enter the atomic age has now arrived' and laying out his scheme for a 500-megawatt nuclear power plant to be built at Jervis Bay, 200 kilometres south of Sydney. While the defence benefits of such a reactor were unspoken, there was no mistaking the military potential of the plutonium it would be producing.

The Jervis Bay reactor never got off the ground, although the scheme had strong support from state electricity authorities and planning reached an advanced stage. Detailed specifications were put out to tender and there was broad agreement over a British bid to build a heavy-water reactor. A Cabinet submission was in the pipeline when Gorton lost the confidence of the party room and was replaced by William McMahon, a nuclear sceptic who moved quickly to 'defer' the project.

It would be another 28 years before Gorton finally came clean on the link between the reactor and his ambition for Australia to have nuclear weapons. In 1999 he told the *Sydney Morning Herald* that '[w]e were interested in this thing because it could provide electricity to everybody and . . . if you decided later on, it could make an atomic bomb'. Gorton did not identify who he meant by 'we' (although Philip Baxter was certainly among them) but Gorton and those who shared his nuclear ambitions were unable to win over the doubters in his own government.

Australia signed the non-proliferation treaty in 1970 but even as the government put pen to paper it was clear that Gorton had no intention of ratifying the treaty. Australia would not ratify the NPT until 1973, and then only after McMahon's Coalition government had lost power to Gough Whitlam's Labor Party. As well as ratifying the NPT, the Whitlam government cancelled the Jervis Bay project that had been in limbo since McMahon became prime minister. With the cancellation of the Jervis Bay reactor, Whitlam effectively ended Australia's quixotic bid to become a nuclear power.

Australia never got its own bomb, although as late as 1984 the foreign minister, Bill Hayden, could still speak about Australian nuclear research providing the country with the potential for nuclear weapons. Hayden served as Australia's foreign minister for more than five years before becoming governor-general. His credibility with the left wing of the Labor Party would prove crucial as the Hawke government stared down public opposition to the spy station at Pine Gap.

Chapter 10
Chinese whispers

Leasing chunks of Australia to the Americans for secret military projects had never sat well with elements of the Australian Labor Party. In March 1963, 36 delegates from the Labor Party's federal conference (six from each state) gathered inside a Canberra hotel to debate a request by the United States to build a naval communications station at North West Cape in Western Australia. Labor policy was for a nuclear-free zone in the southern hemisphere; this could not easily be reconciled with hosting a US communications station at North West Cape to direct the navy's nuclear-armed submarines. The ALP leader, Arthur Calwell, and his deputy, Gough Whitlam, were anxious not to be seen as hostile to the American alliance and hoped that the special meeting of the conference would rescue them by voting in favour of North West Cape.

Just before midnight on Tuesday, 19 March, with the delegates preparing to cast their votes, Calwell and Whitlam were photographed conferring with Labor apparatchiks on the pavement outside the Hotel Kingston. The vote went in Calwell's favour, but on Friday morning Sydney's *Daily Telegraph* published photographs of Labor's parliamentary leader and deputy leader standing anxiously outside while the special conference decided the party's policy.

Two weeks later a Liberal Party backbencher, Harry Turner, coined the damning phrase '36 faceless men' to describe the delegates—ignoring the fact that one of the delegates was a woman. The prime minister, Robert Menzies, borrowed the phrase and used it to flog the Labor Party up and down the country in the run-up to the 1963 federal election. While Calwell won the policy battle over North West Cape, Labor was destroyed in the November election.

The party structure had changed since 1963, but the ALP leadership was still haunted by the scene outside the Hotel Kingston. The party had tied itself in knots over North West Cape and would tie itself in knots again over Pine Gap. In 1969 Harry Turner was still baiting Labor from the safety of the government backbenches:

> What is Labor policy? First of all it is determined from year to year by thirty-six faceless or double-faced men. So we never know what the policy is from year to year . . . They are anti-American . . . They are pro-Marxist . . . A lot of them are pacifists which is a very noble thing, morally speaking, but not a defence policy.

It was not only Australia's conservative government that worried about the Labor Party's loyalty to the American alliance. The CIA was worried too. The agency did not trust Gorton to keep Australian troops in Southeast Asia but it trusted the Labor Party even less.

By October 1969 the government's waning popularity had given Labor a serious chance of winning the impending federal election. The daily intelligence briefs given to the US president reveal CIA concerns that an incoming Labor government led by Gough Whitlam would not only pull Australian troops out of Southeast Asia but would fundamentally rethink the nature of the US–Australia alliance.

> For the President only
> Top Secret
> 18 October 1969
> Prime Minister Gorton's coalition has been losing ground in public opinion polls since August, and next Saturday's election could go either way. Gorton, however, has resumed the outspoken, no-holds-barred political style that has served him so well during most of the past twenty years. As always, local issues and personalities will affect the outcome—factors that are generally unpredictable.
>
> The continuation of Australia's present foreign policy is perhaps the major issue in the campaign. Gorton is forcefully defending the allied presence in Vietnam and the 'forward defense' policy in Malaysia and Singapore. If the Labor Party takes over we can expect a reduction—if not the complete withdrawal—of Australia's contribution to the defense of Southeast Asia. We can also expect a more querulous approach to US-Australian relations and a more questioning attitude towards US activities that affect Australia.

In the election Whitlam's Labor won a slender majority of the two-party preferred vote and became the biggest single party in the House of Representatives, but fell four seats short of tipping Gorton out of government.

The parliamentary stand-off over Australia's hosting of US spy stations continued. Opposition MPs resented the government's restrictions on what they were allowed to know about Pine Gap. Ministers went on refusing to answer questions on operational matters.

Bill Hayden, who would lead the Labor Party after Whitlam, saw straight through the fiction that Pine Gap was only for 'experimental research', declaring that the station was 'quite obviously' being used in association with US spy satellites. After that, however, it was all guesswork: 'A lot of these satellites are designed to track across the Asian land mass—to track across China where, from 100 miles in space, they are able to discover what sort of crops are being sown, whether the crops are diseased and the likely yield at the end of the season. They are able to track troop movements. They are able to report on all sorts of installations. They are able to report on shipping movements and a whole range of other activities.'

Hayden was right about the spy satellites, but it was not Chinese crops they were interested in.

While the government stonewalled, a steady trickle of facts and well-informed speculation about Pine Gap continued to find its way into foreign publications. McMahon's ousting of Gorton as prime minister did nothing to lift the veil of secrecy. For a few months Gorton held the role of defence minister before McMahon replaced him with David Fairbairn. The new incumbent quickly signed up to the old policy of obstruction and evasion. 'Members of the Government with a "need to know" are fully conversant with the activities of the facility,' Fairbairn told parliament after just a month in the job. 'Some Australian officers with the need to know have full access to all areas of the facility. Federal and Northern Territory members of Parliament may enter with prior permission but certain areas are restricted to those with a need to know ... Certain people both inside and outside this House ask why there should be any security at all there. The result of this questioning is constant attempts to penetrate security at Pine Gap by sections of the media and certain honourable members opposite ... I have no intention of either confirming or denying speculation about the purposes of the joint space research facility in Australia.'

A month later, in October 1971, questions were asked in parliament after the authoritative British journal *Jane's All the World's Aircraft* reported that US satellite photographs of Russian and Chinese missile tests were being relayed through a 'secret' US base in Australia. The base in question was identified as Nurrungar but the article reported that Nurrungar was 'controlled' from Pine Gap.

This was not the first time China had been identified as a likely target for spy satellites operated from Pine Gap but it happened at an awkward moment. Just three months earlier, the Labor leader, Gough Whitlam, had made a groundbreaking visit to China. His conversation with the Chinese premier, Zhou Enlai, in the Great Hall of the People had been reported around the world. The pair had spoken in some detail about Australia's military alliance with the United States. Since then, President Nixon had announced his intention to visit Beijing, and the People's Republic of China was poised to replace the Republic of China (now Taiwan) at the United Nations. As a result of the report in *Jane's All the World's Aircraft*, the McMahon government now found itself accused in the Senate of spying on 'another friendly country'.

McMahon had been entirely blindsided by the international reaction to Whitlam's visit to China. Shaking hands with Zhou Enlai in the Great Hall of the People was the symbolic culmination of a personal campaign by Whitlam to recast Australia's relationship with China, tainted as it was by the 'White Australia' policy and by talk of 'yellow' (Asian) and 'red' (communist) perils. The wily Zhou Enlai, however, was after more than just a handshake. According to Billy Griffiths' account of the meeting in *The China Breakthrough: Whitlam in the Middle Kingdom, 1971*:

> Zhou's main goal was to draw Whitlam into denouncing Australia's alliance with America under the ANZUS treaty. Several times he manoeuvred the discussion so that the two men found an area of agreement, then he would passionately assert China's view and pause to hear Whitlam's own—daring him to disagree.
>
> On one occasion, the premier drew a comparison between Australia's relationship with America and China's pact with Russia . . . In recent years, China's dealings with its ally, the Soviet Union, had turned sour through mutual suspicion and doctrinal divergences. Zhou still felt betrayed and he warned Whitlam against trusting unreliable allies, asking, 'Is your ally very reliable?' Whitlam was careful to reject the parallel: there had been no similar deterioration in relations between Australia and America. Zhou Enlai threw up his arms. 'But they both want to control others.'

Lurking in the background of Whitlam's conversation with Zhou Enlai was the issue of nuclear weapons, which China possessed and Australia did not. It was only a quarter of a century since the end of the Second World War and both countries remained wary of what Whitlam referred to as 'Japanese militarism'. There was, said Whitlam, 'one thing about Japan that we do appreciate. It is the most wealthy and developed country which will not have anything to do with nuclear weapons. We think that is reassuring.'

Zhou was less convinced about this than Whitlam. While acknowledging that 'all Japanese people of course do not want nuclear weapons', Zhou suggested that 'a proportion' of those 'in power' were keen for Japan to have nuclear weapons.

What is more, Zhou went on, 'the American Department of Defense is considering whether to give them tactical nuclear weapons or something more powerful'. In response to Whitlam's assertion that acquiring nuclear weapons 'would be in breach of their treaties', the Chinese premier replied:

Behind all treaties there are secret treaties. Otherwise why is it that many secret documents are being published? There are even more secret documents that are not being published. That is why the world is not tranquil.

In Chinese minds, the US–Australian military alliance was a manifestation of an ongoing hostility towards China. Robert Menzies had explicitly linked the two in April 1965 when he justified sending Australian combat troops to Vietnam by asserting that the Vietnam War was 'part of the downward thrust of China between the Indian and Pacific Oceans'. Menzies' successor, Harold Holt, further alienated China in 1967 with his bizarre decision to open an Australian Embassy in Taipei, capital of the Republic of China. (Menzies had always maintained that there should be no Australian ambassador in Taipei.)

This decision, described by Whitlam as 'one of the oddest episodes in our diplomatic history', made it impossible for the government of William McMahon to recognise China on terms that the Chinese could ever accept.

But Australia's Vietnam commitment was only one facet of a security alliance that also saw Australia hosting US spy stations that were intercepting Chinese telemetry signals and eavesdropping on Chinese communications. Zhou's observation that 'behind all treaties there are secret treaties' summed up the backroom deals and hidden agreements that lay behind the Joint Defence Space Research Facility at Pine Gap.

Whitlam returned from his privately funded Chinese trip to predictable outrage and ridicule from McMahon, who famously remarked, 'In no time at all Zhou Enlai had Mr Whitlam on a hook and he played him as a fisherman plays a trout.'

In fact it was McMahon who landed in the net. Just days after Whitlam's visit, Zhou Enlai received a secret visitor: President Nixon's special envoy, Henry Kissinger. Zhou had personally fixed the dates to ensure that Whitlam came first. As a result of Kissinger's visit, Nixon announced that he would travel to Beijing early in 1972 to seek the normalisation of diplomatic relations. Characterising his visit as a 'journey for peace', the US president declared his 'profound conviction that all nations will gain from a reduction of tensions and a better relationship between the United States and the People's Republic of China'.

Whitlam, who in 1954 had been the first Australian member of parliament to urge recognition of the People's Republic of China, had taken a big political risk in visiting China, but the gamble paid off when it became clear to the world that Nixon was following the same path. Rather than being out on a limb, Whitlam found himself at the epicentre of what Billy Griffiths calls 'a seismic shift in America–China relations'. As Bill Hayden would later recall, a disaster in the making for Whitlam was transformed into 'a stroke of genius'.

The radical change in US policy towards China that quickly followed Whitlam's visit left McMahon fuming. The Americans had alerted him just hours ahead of Nixon's announcement. McMahon complained bitterly about not having been forewarned. The lack of discussion made a mockery of the supposedly close relationship between the US and Australia that was at the heart of the ANZUS alliance and was the basis for Pine Gap and other 'joint' military facilities in Australia. McMahon tried to play down the significance of Whitlam's China visit, and to play up the fact that Zhou had considered it unnecessary to inform Whitlam of Kissinger's impending visit. To Whitlam, however, the fact that the US president was following in his tracks was a complete vindication of the risks he had taken.

As opposition leader, Whitlam had run rings around McMahon on China, but his election as prime minister was more than a year away. In the meantime, the Coalition government still held the keys to Pine Gap.

Chapter 11
Keep your hands off

By the start of 1969 the Joint Defence Space Research Facility was operational, but there was still no sign of an Australian technician in the operations building. In March 1969 Jim Bullen, the grazier on whose land Pine Gap had been built, told the ABC's *Four Corners*, 'I don't think you would find, except for the Commonwealth Police... more than perhaps one man wheeling a barrow that's an Australian. There might be more... but there wouldn't be many more.' Bullen, of course, was reputed to have insisted when he sold the land that the new facility be a 'wholly Australian' project. It did not sound like the sort of thing a Northern Territory sheep farmer would say, but if it was true then Bullen could count himself among the many Australians who felt deceived by the government's promises on Pine Gap.

It was not until November 1970 that the first Australian computer operators were allowed inside the signals processing station. This was one of the three sections in the operations building; the other sections were station-keeping (which controlled the orbits of the satellites and kept their listening antennas pointed in the right direction) and signals analysis, which analysed the intercepted material. Signals analysis was the facility's most highly classified section and no Australians were permitted inside.

According to Des Ball, Bill Robinson and Richard Tanter, in 1970 there were 440 personnel employed at Pine Gap, 184 Australians and 256 Americans. By now Australians were supposed to comprise half of the staff at Pine Gap but 'few of the Australians were allowed into the . . . Operations Building'. In October 1970 the *Northern Territory News* reported that there were just nine 'Australian scientists and technologists' among the Australians employed at Pine Gap.

One of the four Australian computer operators who arrived in November was Leonce Kealy. Before working at Pine Gap Kealy had spent nine years with the Royal Australian Air Force. He had left the air force and was working on computers for the defence department when he saw a job ad for computer programmers. According to Kealy, all four of those who went to Pine Gap were from the defence department. His security clearance took 'several months' to come through.

One of few insiders to have spoken about his time at Pine Gap, Kealy soon noticed the imbalance in operational staff numbers inside the supposedly 'joint' facility. In his memoir, *The Pine Gap Saga: Personal experience working with the American CIA in Australia*, Kealy wrote:

> What the Americans did was to make a huge list of all personnel at the base, including those in the unclassified outer perimeters, who included housemaids for the hotel units, cooks, gardeners, laborers, bus drivers and clerical

staff. This allowed the Americans to satisfy the 50/50 relationship admirably, but leaving almost entirely all Americans in the Top Secret sector.

Interviewed in 1977 by the *Canberra Times*, Kealy recalled ruefully, 'The Americans run that place and they think they run Alice Springs. There has been a lot of drivel written about Pine Gap but one thing is for sure. It is not meant to be a place where Australians can feel comfortable.'

Jim Pidcock was another of the Australian computer operators who arrived with Kealy from the Department of Defence. Pidcock told Kealy in 1975:

> There's no doubt... that they must have been forced, politically, to employ Australians and have spent every minute since trying to get rid of them. There is no other possible explanation... I'll never forgive the bastards; every time I hear an American accent it still gives me pain.

In 1970 the staff in the signals processing station were organised in four daily shifts. With one Australian rostered on per shift, it is not surprising that they all felt marginalised. Seven years later, Kealy told the *Canberra Times* that his first reaction on entering the central computer complex was one of 'awe'. While still forbidden to reveal information about operational equipment (such as the number of computers), Kealy did mention that the computer room was so big that 'headphones were needed to communicate from one end of the room to the other'.

A fifth Australian computer operator arrived at Pine Gap in April 1974, and by February 1975 nine more had been recruited.

According to Des Ball, none of the fourteen Australians was employed in a supervisory position; they felt there was 'a deliberate policy to exclude Australians from these positions'.

In his book, Kealy complained that American technical staff were often promoted over the heads of Australians with higher qualifications and more on-the-job experience. Like their American colleagues, the Australian computer programmers worked not for the Australian government but for the US contractor E-Systems Inc., based in Dallas, Texas. One of three major technical contractors at Pine Gap (the other two were IBM and TRW Inc.), E-Systems managed the main computer room. In 1973 the company advertised a new position of lead programmer analyst to supervise the computer room at Pine Gap. The job was advertised in Dallas but not in Australian newspapers. By the time Kealy heard about it, the position had been filled by an American from outside the complex.

Believing he was qualified, Kealy protested to management and the union. He claims to have been warned afterwards by 'several American friends' to 'watch your back for the next day or two'. That night Kealy 'came into work to find a nasty message in the log book . . . It read, "Kealy keep your hands off all programming in future".'

During his years at Pine Gap Kealy had written some valuable programs, including one that enabled operators to track and intercept signals from a particular Soviet satellite as it passed over the station. Embittered by what he considered to be discrimination against himself and other Australians working at Pine Gap, Kealy sat down in front of his screen on the morning of 6 February 1975 and wiped his programs from Pine Gap's computers. Accused of 'sabotage', he was dismissed by E-Systems 24 hours later.

Others shared Kealy's belief that Australians were getting a raw deal at Pine Gap—and not just over staffing ratios. Suspicions were growing that Australia was providing 'real estate' for a secret and dangerous US military project that might have been in America's national interest but was not in Australia's.

In the wake of Kealy's removal, a letter written on behalf of the minister for defence stated that staffing ratios and the selection of Australians for senior positions were issues that Canberra would be taking up 'as a matter of principle, not with the firm [E-Systems Inc.] but with the senior United States Government representative who will direct the firm accordingly'. The writer conceded, however, that the task of balancing the numbers of Australian and US employees and ensuring some Australians were hired for senior positions was 'a slow process'.

The staffing situation in the operations building would scarcely improve in the decade following the arrival of the first four Australians in 1970. Eight years later Des Ball wrote that of the '226 Australians employed at Pine Gap [in 1978], only about 16 are involved in performing technical functions'.

Chapter 12
The birds

Leonce Kealy and his Australian colleagues arrived at Pine Gap just in time to receive the data that had begun streaming from the CIA's first Rhyolite satellite. The satellite—dubbed 'Bird 1' by Pine Gap staff—had been launched in June 1970 from Cape Canaveral in Florida and had spent several months over the United States undergoing performance tests by the manufacturer, TRW, and engineers from the CIA's Directorate of Science and Technology. These tests confirmed that the satellite, as well as being able to pick up telemetry signals, was highly sensitive to communications transmitted at VHF, UHF and microwave frequency bands. In his book *The Falcon and the Snowman*, Robert Lindsey writes that the Rhyolite satellites 'carried a battery of antennas capable of sucking foreign microwave signals from out of space like a vacuum cleaner picking up specks of dust from a carpet: American intelligence agents could monitor Communist microwave radio and long-distance telephone traffic over much of the European landmass, eavesdropping on a Soviet commissar in Moscow talking to his mistress in Yalta or on a general talking to his lieutenants'.

For more than 30 months Bird 1 was America's only operational geosynchronous SIGINT satellite (a second Rhyolite satellite, launched in December 1971, failed to reach orbit after a booster malfunction). During this time it was repositioned several times to monitor places of interest to its CIA handlers. According to Jeffery Richelson, its first assignment was 'probably... over Borneo to receive telemetry signals from liquid-fuelled ICBMs launched from Tyuratam in a north-easterly direction toward the Kamchatka Peninsula impact zone'. Des Ball suggests that in 1971 it was repositioned a number of times to monitor the India–Pakistan war and the Vietnam War.

The second successful Rhyolite satellite was launched in December 1972 and the third in March 1973. Bird 2 was positioned to intercept Soviet signals while Bird 1 eavesdropped on China and Vietnam. According to James Bamford's book *The Puzzle Palace: Inside the National Security Agency, America's most secret intelligence organization*, the third bird was 'parked' above the Horn of Africa from where it could listen in on microwave transmissions from western Russia as well as intercept telemetry signals transmitted from missiles launched from both Tyuratam and Plesesk.

None of this was shared with the Australian public or even the parliament, despite the fact that the early birds were controlled from Pine Gap. Inquisitive Australian MPs were, however, told that '[t]he grounds of the Facility, which are extremely well kept, are being improved by a programme of planting lawns, trees and shrubs, including native species suited to the harsh climatic conditions'.

Questioned by Australian reporters about whether Pine Gap could now be considered a nuclear target, the visiting US assistant defence secretary, David Packard, replied that in the event of a nuclear war 'Pine Gap would be one of the places I'd like to be'.

The COMINT (communications intelligence) function of the Rhyolites—the interception of radio and other electromagnetic transmissions—was aimed mainly at the USSR, China and Vietnam, although they continued to listen in on communications from Indonesia, Pakistan, India and Lebanon. A later generation of geosynchronous satellites, controlled not from Pine Gap but from its British sister station at Menwith Hill in Yorkshire, was specially designed to target communications from the western USSR, Eastern Europe, the Middle East and the Mediterranean.

The early Rhyolite satellites revolutionised the collection of signals intelligence. In his book *UK Eyes Alpha: Inside story of British intelligence*, Mark Urban writes that the CIA's new 'birds' could 'pick up various types of transmission, but the most important take came from the microwave telecommunications links which by the 1970s had been installed across the Soviet Union. Microwave circuits had been considered highly secure by the Russians because they use a narrow beam of energy between a transmitter and receiver which have to be within line of sight. Trying to pick up the microwave transmission from even a few miles away is pointless. But the parts of the microwave beam which shoot past the receiver—spillage in [SIGINT] jargon—continue in a straight line up into space. Microwave beams may also strike the ground in places, bouncing signals straight upwards. RHYOLITE was . . . able to "hover" over the Soviet Union. It was equipped with a large parabolic dish so that the feeble fragments of microwave energy could be refocused on its receiver. Each microwave circuit could carry hundreds of conversations. The possibilities of RHYOLITE were, says one [SIGINT] insider, "mind-blowing".'

A former signals intelligence officer told Urban, 'When RHYOLITE came in, the take was so enormous that there was no way of handling it.' Billions of dollars were spent on systems capable of receiving, processing and analysing the data.

An upgraded satellite, called Argus or AR (for Advanced Rhyolite), was launched from Cape Canaveral on 18 June 1975 aboard a Titan rocket.

In his book on Pine Gap, Des Ball cites British estimates that the Argus was twice as big and twice as heavy as the Rhyolite, with a much larger antenna. The greatest operational improvement was in its ability to eavesdrop on microwave communications, including voice transmissions. Ball quotes a British newspaper report saying the Argus 'could listen in to any part of the [Soviet] microwave network and thus eavesdrop on discussions—for example—between members of the Politburo or Soviet military commanders. It could also intercept short-wave radio communications between, say, Soviet tank commanders on the Polish border. It could even penetrate the [data links between] Soviet military computer systems.'

More capable than the Rhyolite, but much more expensive, the Argus was favoured by the CIA but opposed on cost grounds by nearly everyone else. Only one Argus satellite was launched before Congress vetoed the program.

Chapter 13
Number two on the shit list

On 2 December 1972, after five and a half years as Labor leader, Gough Whitlam defeated William McMahon at the polls and was elected prime minister of Australia. The 36 months of the Whitlam government, from December 1972 to November 1975, would plunge the US–Australia alliance into its deepest ever crisis. Although Whitlam and President Nixon were following a similar course on China (Whitlam told the Australian ambassador to Japan that he was 'glad to be a pathfinder for Nixon') and on détente with the Soviet Union, the Australian Labor Party was strongly tainted with anti-Americanism. In Canberra and Washington there were fears about Labor's commitment to the American alliance.

By declaring his intention to visit China just days after Whitlam's meeting with Zhou Enlai, Nixon delivered the Labor leader a diplomatic and political coup that few in Australia (and certainly not Prime Minister McMahon) had foreseen. But Whitlam's comments to Zhou, heard and reported by the Australian press, carried a sting for the US president. 'The American people have broken President Lyndon Baines Johnson,' Whitlam told the Chinese premier, 'and if Richard Milhous Nixon does not continue to withdraw his forces from Vietnam they will destroy him similarly.' While resisting Zhou's attempts to draw him into directly criticising the United States, Whitlam declared that 'throughout the north-east [United States]' there was 'no support among opinion makers' for US conduct of the war. 'The American people will force the American president to change the policy.'

Whitlam's prophecies 'carried intent', James Curran writes in *Unholy Fury: Whitlam and Nixon at war*. '[H]e was not only warning Nixon about his potential political demise, but also playing up the influence of protest and dissent on the administration's policy.'

The deputy chief of mission at the US Embassy in Canberra, Hugh Appling, described Whitlam's comments on American politics as 'meddlesome'. The notoriously foul-mouthed Nixon probably had a different description.

As the Australian Labor Party began to scent victory in the 1972 federal election, its questioning of the US–Australian alliance grew louder. American diplomats were 'distinctly wary of this new breeze in Australian politics', Curran writes. 'Already frustrated by John Gorton's unpredictability and ... William McMahon's irascibility, they wondered what kind of future the alliance would have under a Labor government.'

The signs looked anything but positive. At the ALP's 1971 annual conference in Launceston there had been strong moves by the party's left wing to end the ANZUS Treaty and to adopt a policy of 'non-alignment'. The bid was only thwarted when Whitlam, the party leader, declared it would be 'intolerable' for him to have to oppose ANZUS on behalf of the party.

Still, the conference accepted unanimously a proposal that the Labor Party should delete from its platform a clause that said 'the alliance with the US and New Zealand is essential and must continue'. In its place went a statement that the Australian Labor Party 'seeks close and continuing co-operation with the people of the United States and New Zealand to make the ANZUS Treaty an instrument for justice and peace and political, social and economic advancement in the Pacific'. The new ALP policy on ANZUS was reported to have been 'drafted by Mr Whitlam'.

Australia's value to the United States went far beyond the secret bases. As the quarterly US journal *Foreign Policy* noted in 1982, Australia was 'one of the most strategically valuable pieces of real estate on the planet . . . 69 per cent of Japan's oil requirements, 70–80 per cent of Western Europe's and 15 per cent of America's passes through the area between Australia and southern Africa. US B-52s flying from Guam to Diego Garcia refuel in Australia . . . Australia hosts 10 American military installations. Because of their unique location, most cannot be replicated at any cost. The new US Defense Guidance characterizes Australia as a critical area.'

On the question of US military bases, Whitlam had previously taken the position that Labor would not oppose them provided the electorate was properly informed about their function. At the ALP's 1969 conference in South Australia, Whitlam had managed to avoid committing the party to any definitive statements about Pine Gap. Whitlam's guarded pronouncements were at odds with the more outspoken comments of left-wingers such as Jim Cairns, who accused Gorton and his Coalition predecessors of having 'made Australia a nuclear target by stealth' by agreeing to host US military facilities such as those at Pine Gap and North West Cape.

In November 1972 Nixon's national security advisor, Henry Kissinger, had bigger problems on his hands than an election in Australia. In Paris, he was trying to sell 'peace with honour' to the Vietnamese, while at home Nixon's decision to mine North Vietnamese harbours had provoked the most intense anti-war protests since US troops invaded Cambodia in 1970. Chile's Marxist president, Salvador Allende, was nationalising US-owned copper mines and in the Middle East Egypt's President Anwar Sadat was dragging the Arab nations towards war with Israel.

Under a string of conservative leaders, Australia had proved a staunch and subservient ally. Not only had Harold Holt vowed to go 'all the way with LBJ' but John Gorton had promised Nixon, at a White House dinner in 1969, 'sir, we will go Waltzing Matilda with you'. By the time the war was over, nearly 60,000 Australian troops had served in Vietnam and more than 500 Australians had died. US predictions that a more sceptical and less slavish Australian Labor Party would win the 1969 federal election had been proved wrong. The same predictions were being made more urgently in the lead-up to the 1972 election, and this time the odds of a Labor victory were much shorter.

Labor's 'It's Time' campaign slogan, coupled with a progressive policy program and Whitlam's personal charisma, galvanised an electorate disillusioned by Vietnam and ready for a change after 23 years of Coalition rule. In the end, the judgement of the American Embassy in Canberra was that Whitlam would lose the election, but when the results came in Labor had scraped into government with a majority of nine seats.

The CIA's advice to the president before the 1969 election—that an incoming Labor government would likely have 'a more querulous approach to US-Australian relations and a more questioning attitude towards US activities that affect Australia'—was even more pertinent now.

Whitlam was immediately under pressure to reassure the Americans that a Labor government could be trusted to look after their interests—in particular the maintenance of US bases in Australia. Two days after the election, the US president's top-secret daily brief included several paragraphs about Australia:

> Prime Minister-elect Whitlam made it clear over the weekend that the alliance with the US will remain the cornerstone of Australian foreign policy, but Labor Party spokesmen have promised to re-examine the agreements permitting US military and scientific installations in Australia.

This straightforward reporting of events in Australia was followed by two paragraphs of CIA analysis:

> It is clear that some party leaders, including prospective defense minister Barnard, are not well versed on the US facilities. Once they become fully informed on the benefits Australia derives from the installations, we believe the new government will conclude that they fit within the framework of the ANZUS treaty, the formal basis of the US alliance. Whitlam may seek a more independent position within the alliance, however.
>
> The new administration will make some changes in the previous government's 'forward defense' doctrine. Most immediately, it is likely to bring home the small training contingent left in Vietnam after the withdrawal of Australian

combat forces last year. Whitlam has said that Australia will honor its commitments to Malaysia and Singapore under the five-power agreement with the US and New Zealand, but wants eventually to reduce the Australian military presence in Malaysia and Singapore to training and assistance missions arranged bilaterally.

The CIA, of course, had a vested interest in the new government's policy towards Pine Gap. The first chief of facility at Pine Gap, Richard Lee Stallings, and his successor, Harry Elmer Fitzwater, were both CIA men. CIA technicians controlled the spy satellites. The agency had expended vast quantities of money, manpower and equipment on developing and running the place. Pine Gap was an investment that could not be allowed to fail.

While Whitlam awaited the results of the Senate election, he arranged with the governor-general for himself and Lance Barnard, the deputy prime minister and minister for defence, to share all the portfolios between them. Once the senate results came through, the Labor caucus would be able to elect a full ministry.

Within hours of Whitlam winning office, Barnard was on the phone to Arthur Tange, a former high commissioner to India who had returned to Canberra to become secretary of the Department of Defence. As the top public servant in the defence department, Tange had responsibility for both Australia's military intelligence and its signals intelligence. Barnard was keen for Tange to remain in the job.

Tange's first impressions of the Labor government were positive. In his memoir he contrasts the vigour of the new administration with the inertia of its predecessor. 'In contrast,' he writes, 'with the vacuum in policy-making and the politically defensive attention to trivia that had characterised the last few months of the McMahon Government, Labor deluged us with policy objectives and organisational changes.'

Renowned, like Whitlam, for his intellect, Tange felt he understood the new government's defence policies, some of which he had advocated, without success, to their Coalition predecessors. He had qualms, however, about Labor's attitude to the US–Australian alliance. In particular he was concerned about Whitlam's ability to handle the party's anti-American faction, which had long had Pine Gap in its sights. In his book Tange recalls that he 'could not know Labor's intentions, as distinct from rhetoric, in respect of the American connection. I was uncertain as to how the Ministers in office would handle publicly declared policies of disclosing, or possibly terminating, the activities of the United States in the facilities shared with Australia at Pine Gap and Nurrungar. I hoped to dissuade the Ministers from either course when they were given the facts to which . . . they had not been made privy when in Opposition.'

Before Tange or anyone else had time to see how Labor's 'intentions' departed from its 'rhetoric', the American connection was thrown into crisis by President Nixon's decision to bomb Hanoi and Haiphong. The raids came after North Vietnamese negotiators had broken off peace talks in Paris. Nixon and Kissinger suspected the North Vietnamese of playing for time. The new Democratic-dominated 93rd Congress would go into session on 3 January 1973 and Nixon feared the incoming US Congress would simply cut off funds needed to fight the war, denying him the chance to deliver on his promise of 'peace with honour'. The scale of the Hanoi raids, which began a week before Christmas, was supposed to drive the North Vietnamese back to the negotiating table.

The campaign began on 18 December 1972 with more than one hundred B-52 bombers taking off from US bases in Guam and Thailand. Bombing was suspended on Christmas Day but the attacks resumed the next day before coming to an end on 29 December. More than 1600 North Vietnamese civilians were reported to have been killed.

The Christmas bombing of Hanoi made no significant difference to the North Vietnamese negotiating position, with both sides finally agreeing to terms that had been largely accepted in October. If anything, the compromises lay on the American side, prompting a Kissinger aide, John Negroponte, to remark, 'We bombed them into accepting our concessions.'

Three days after the first raid, Whitlam sent a letter of protest to Nixon via the Australian ambassador, James Plimsoll. In *Unholy Fury*, James Curran records that Plimsoll was 'not exactly an enthusiastic envoy' and that he confided in Marshall Green, the former US ambassador to Indonesia who was now assistant secretary of state for East Asia and the Pacific, that he regretted Whitlam's failure to discuss the matter with him before sending the letter. Plimsoll felt Whitlam had 'acted under pressure from other Cabinet members'.

The text had been drafted by Sir Keith Waller, a former ambassador to Washington, who felt the letter was 'reasonably moderate'. But that was not how Nixon read it.

Whitlam stated that the collapse of the Paris peace talks had come as a 'bitter blow' to him, the government and 'the Australian people as a whole' and questioned whether the bombing of Hanoi (like previous US escalations of the war) would have the effect of making the North Vietnamese more rather than less intransigent. Not wanting his letter to be taken the wrong way, Whitlam then expressed his hopes for 'a period of positive cooperation between our two countries on a wide range of matters'. So far, perhaps, so 'moderate'. But what came next undid all his careful conciliation. Whitlam outlined his intention to invite other East Asian nations to join Australia in 'addressing a public appeal to both the United States and North Vietnam to return to serious negotiations' for peace.

Nixon was livid, not only at the reproof but at the implication that democratic Washington and communist Hanoi were somehow morally equivalent. While Whitlam did not intend to release the text of his letter to Nixon, he signalled that he would make it known to the Australian people that a letter had been sent. In Whitlam's gargantuan account of his three years in power, he confirms that the government did disclose that a 'stiff message' had been delivered to the president.

When the letter was shown to Kissinger, he immediately called Plimsoll at the Australian embassy, only to be told that Plimsoll had left Washington and was en route to Australia for consultations with his prime minister. Plimsoll's number two, Roy Fernandez, took the ear-bashing intended for his boss. A furious Kissinger railed against Whitlam for having put America 'on the same level as our enemy' and warned Fernandez what would happen if the prime minister's idea of a public appeal to both sides ever got out. Fernandez was left in no doubt that the alliance itself was in jeopardy. According to Curran, Kissinger's call was 'so swift and so brutal that the White House staffer who typed up the verbatim transcript misspelt the Australian capital as "Kenbrook" and Sir Keith Waller, an eminence grise of Australian diplomacy, made his only appearance in the document as "skeef walla"'.

Whitlam's measured criticism of Nixon's handling of the war was almost benign by comparison with what came after.

Around the world, the Hanoi raids were condemned. After the Swedish prime minister compared them with Nazi massacres of Jews and the bombing of Guernica, Nixon broke off diplomatic relations. Many of Whitlam's ministers were similarly revolted.

Jim Cairns, now Whitlam's minister for trade, described the Hanoi bombings as 'the most brutal, indiscriminate slaughter of defenceless men, women and children in living memory'. At the same time the Seamen's Union slapped a black ban on all US ships arriving in Australia. The next day Whitlam's minister for employment, Clyde Cameron, called on all nations to isolate the United States 'commercially and diplomatically until Congress moves in to control the maniacs who seem to be determining US policy in Vietnam'. Tom Uren, minister for urban affairs, went even further, denouncing Nixon personally for his 'hypocrisy', 'arrogance' and 'double dealing' and describing his Vietnam policy as 'mass murder'.

Marshall Green would later tell the *Sydney Morning Herald* that 'President Nixon was furious. He elevated Australia to number two place on his personal shit list. Sweden occupied number one place. But Australia stayed number two for quite a while... You certainly couldn't mention Whitlam's name around the White House without an explosion.'

Eventually Whitlam slapped down Uren's comments as being 'grotesquely offensive' and dismissed threats by Australian unions to boycott all American interests as 'so much crap'. However, he refused to resile from his criticism of the Hanoi raids. At a meeting with the US ambassador, Walter Rice, in early January, Whitlam threatened to bring in the United Nations if the US suspended peace talks or resumed its bombing of North Vietnam. While Whitlam had long supported the presence of US intelligence installations, he also cautioned Rice that 'if there were any attempt to screw us or bounce us, inevitably these defence arrangements would become a matter of contention here'.

It was not a private conversation. Whitlam expected his words to the ambassador to be reported back to Washington. The message—or warning—to the White House from Australia's new prime minister was that support for its intelligence facilities was not, after all, unconditional. What had been given could also be taken away.

Or could it?

Chapter 14
'I can't stand that [****]'

The day after he won office, Whitlam introduced his political staff to the heads of the public service. At the meeting the newly elected prime minister made a startling announcement. According to his speechwriter, Graham Freudenberg, Whitlam said, 'I take full and personal responsibility for these men. I vouch for them and they are not to be subject to security clearances. These men are not to be harassed by ASIO.' None of the mandarins demurred, although another of Whitlam's speechwriters, Evan Williams, noted later that ASIO had probably already sought and obtained whatever clearances it wanted.

It was a fortnight before the ramifications of Whitlam's swashbuckling assault on the government's security apparatus began to be canvassed in public. On 19 December a short article appeared on page seven of the *Canberra Times* under the headline 'ASIO CLEARANCES: Decision by PM creates problems'. Whitlam's statement was expected to cause 'immediate difficulties with US and British intelligence agencies with whom ASIO co-operates directly'. According to the article's author, Frank Cranston:

> Defence, as well as political intelligence areas on which Australia has relied heavily in recent years, could quickly be closed unless foreign sources can be guaranteed that confidential information will not be available to people who have not been screened by ASIO.
>
> British and US intelligence agencies accept ASIO security clearances.
>
> Even technical information could be withheld in the absence of assurances of the security integrity of possible recipients.

While government MPs were routinely screened by ASIO, Cranston pointed out that newly elected Labor MPs might not yet have been 'processed'. If any of these were appointed assistants to ministers, their lack of ASIO security clearance could be a major problem.

As secretary of the Department of Defence, Arthur Tange was especially dismayed by Whitlam's decision and continued to use his mandarin cunning to exclude staff without clearances from discussions on intelligence matters.

Within his fiefdom of defence, Tange kept a jealous grip on everything to do with the shared intelligence facilities and the broader US–Australian alliance. He admitted in his memoir to taking a 'highly restrictive view of those entitled to know the unpublished functions of the Facilities'.

Within the Government there had been a strictly limited 'listing' procedure. Most of the Public Service was excluded, including members of my own Department and Foreign Affairs except at the very top. I kept personal control of any discussions with the Americans that developments might make necessary.

Pine Gap was the single most important piece in the US–Australian intelligence jigsaw and Tange guarded its secrets obsessively. He let only one other member of his department, the chief defence scientist, Dr John Farrands, into the mystery of its operations. Both men visited the site on several occasions and were shown its capabilities and the type of data collected. Tange wrote of having been given 'assurances' on his visits to Pine Gap that the installation was not used to spy on Australian targets. He had heard these assurances repeated when he travelled with Australian government ministers for 'high-level meetings' in Washington. Tange believed what he was told and accepted that the data collected by the CIA at Pine Gap 'would be used by the United States to maintain its military capability of matching the Soviet Union's advances'.

A more vivid account of the intelligence being collected from the Rhyolite satellites came from Paul Dibb, who made his first visit to Pine Gap in 1974 as director-general of the National Assessments Staff (the predecessor to the Office of National Assessments). Dibb had previously been a research fellow at the ANU specialising in Soviet affairs and was also working for ASIO, which had tasked him with cultivating Soviet diplomats for intelligence and encouraging defections. As head of the National Assessments Staff, Dibb had top-secret 'absolute' clearances for Pine Gap, Nurrungar and North West Cape. He would hold those clearances for 30 years, including a period when he worked alongside Des Ball at the Australian National University's Strategic and Defence Studies Centre.

Unlike another ANU colleague, Robert Cooksey, Dibb had no need to fly himself up to Alice Springs and loiter outside the fence in the hope of uncovering the secrets of Pine Gap. He was taken straight to the nerve centre, where he was told, 'We are going to patch you through to listen to the Soviet Northern Fleet Commander in Severomorsk talking to naval headquarters in Moscow.' Dibb is cagey in what he reveals about the operations at Pine Gap, pointing out that the writings of colleagues such as Des Ball were viewed by the Department of Defence (and especially Tange) as 'aiding the enemy'. Nevertheless, in his book *Inside the Wilderness of Mirrors*, Dibb describes eavesdropping on the Soviet fleet commander, who 'commanded more than fifty major surface combatants and 130 submarines' including Delta-class submarines armed with twelve intercontinental ballistic missiles. Dibb, author of the 1986 *Review of Australia's Defence Capabilities* and the 1987 Defence White Paper, recalls being 'impressed' with what he heard that day.

Tange's own pronouncements about Pine Gap—such as his insistence that 'activities in Australia were not made targets for information gathering'—tend more to obscure than illuminate. Des Ball has demonstrated that Australia itself is within the 'footprints' of the geosynchronous satellites controlled from Pine Gap. The satellites, as Victor Marchetti observed, sucked up signals 'like a vacuum cleaner'. Signals intercepted by the satellites and beamed back to Pine Gap included microwave communications that could theoretically have included phone conversations and data transmissions, within and to or from Australia. Labor's Gordon Scholes told the parliament on 4 May

1977 that it was 'fairly evident that almost any international telephone call leaving Australia can be monitored by satellite, the information being fed to the United States of America or other countries'.

A former CIA station chief has stated that 'it is not remotely possible to collect personal [Australian] conversations; if by hazard it picked up domestic conversations, it would get no further than the responsible Australian desk officer'. Tange accepted (or at least purported to accept) assurances that Pine Gap did not eavesdrop on Australians. His nemesis, Des Ball, was less sure. 'The question,' Ball wrote in *Pine Gap*, 'is really one of what happens to the recordings of these Australian transmissions once they are sorted out—are they destroyed or passed back unread to the Australian intelligence and security agencies, or are they retained and those of interest read and analysed by the CIA and NSA?'

Whatever idealistic plans Whitlam and Barnard might have had about confiding in the Australian people, Tange's own priorities as head of defence were clear: to convince the new government of the necessity of keeping the US intelligence bases in Australia, and to make sure the purpose of the bases remained secret. There was intense pressure from the left wing of the Labor Party and from the media for Whitlam to deliver on his promise to come clean about what went on at the US installations. Of course, the function of each station was different.

In his memoir Tange refers loosely to the 'Pine Gap/Nurrungar twins'. As the journalist Brian Toohey has pointed out, the CIA-run Pine Gap and the US Air Force tracking station at Nurrungar were not in any real sense twins. The satellites controlled from Nurrungar used an infra-red telescope to detect heat from missile launches. Those operated from Pine Gap were designed to eavesdrop on electronic communications and signals, including missile telemetry.

Honouring Labor's pre-election pledge, Lance Barnard drafted a parliamentary speech explaining the function of the US bases. Whitlam had no intention of shattering the US–Australian security compact and the aim of the speech was not to give away operating secrets but to outline the intelligence-gathering activities of Pine Gap and Nurrungar using information that was already publicly available. When he discovered what Barnard proposed to tell the parliament, Tange hit the roof. According to one of Barnard's aides, Tange insisted that every copy of the proposed speech be recalled.

The speech Barnard eventually delivered to the House of Representatives on 28 February 1973 began by chastising the Gorton government for its refusal to brief Whitlam on the 'activities and functions' of Pine Gap and Nurrungar. Labor, he said, had scrapped the policy of banning visits by members of parliament. Under the new government, MPs would have 'special right of access to the two installations' which would enable them to see 'something' of the nature of their operations. These visits could only be arranged in advance through Barnard himself and would be subject to 'some conditions'. The government might have changed, but there would be no opportunity for left-wing Labor MPs to embarrass the Americans with impromptu visits to the operations room at Pine Gap.

Determined to resist the clamour for 'more information', Tange saw his task as 'preserv[ing] secrecy and . . . reassur[ing] our allies on the point'. As well as guarding against 'wilful breaches of security' by enemy agents and others, Tange was anxious to prevent 'leakages of secrets from carelessness or from people with an urge to parade their unique possession of information'.

According to Whitlam himself, only four ministers were privy to the secrets of the US intelligence stations: himself as minister for foreign affairs, the minister for defence (Barnard), and their assistant ministers, Senators Bishop and Willesee. 'They are entitled to know,' Whitlam told a Melbourne Press Club lunch, 'but they are not entitled to tell.'

Sceptical about government ministers, Tange was downright suspicious of their staff.

> If Ministers had to be informed of developments in writing, I conveyed it personally by hand, declining to use the customary procedure of trusting ministerial staff to place material before a Minister.
>
> My objective was not secrecy for its own sake. It was to prevent the Soviet Union learning about a vital intelligence activity by reading the newspapers simultaneously with the Australian public.

A former school mathematics teacher, Barnard was no match for his devious department secretary. When it came to peeling back the layers of secrecy covering both Pine Gap and Nurrungar, his speech to parliament simply reiterated the mantra of conservative governments by saying that information about the techniques used and the data collected 'must be kept highly secret'. All Tange would allow him to say 'without injury to truth' was that 'neither station is part of a weapons system and neither station can be used to attack any country'. Both statements were disingenuous, since forewarning of an enemy attack was a prerequisite for retaliatory action.

If Tange was not sure he could trust ministers and their staff, he was certain he could not trust the press, who were always 'probing into defence secrets'. He appeared to have little difficulty persuading Whitlam to adopt a policy 'neither to confirm nor deny' media speculation about what the Americans were up to at Pine Gap and Nurrungar.

The reason Whitlam always gave for not revealing what went on at the 'joint' bases was that 'they are not Australian secrets'. The communications station at North West Cape in Western Australia was not a joint base. It was run by the US Navy. At the same time as his defence minister was bamboozling the parliament on Pine Gap, Whitlam launched a scathing attack on the 'thoroughly obnoxious' agreement that ceded total control of North West Cape to the US government for 25 years. He told his audience at the Press Club that no 'self-respecting' government would have entered into the agreement and promised that Barnard would renegotiate the deal 'in a few months'.

Unnerved by the Whitlam government's decision to dispense with security clearances and smarting from its criticism of the Hanoi bombing, the Nixon White House now had more evidence of Australian backsliding. If the Labor government was determined to 'renegotiate' the deal on North West Cape, what were its plans for the joint intelligence facilities? The lease on Pine Gap was due to expire in 1976; the Nurrungar lease three years later. The CIA watched anxiously for further signs of instability in Canberra. It did not have long to wait.

At eight o'clock on the morning of 16 March 1973, uniformed Commonwealth Police led by the attorney-general, Lionel Murphy, barged into the Melbourne headquarters of the Australian Security Intelligence Organisation (ASIO). Murphy had been warned by the New South Wales police that

Croatian fascists might try to assassinate the Yugoslav prime minister while he was visiting Australia. At the time, ASIO was mainly concerned with Soviet espionage and left-wing subversion. The attorney-general suspected the organisation of protecting the Croatians by withholding information. Striding through the building, Murphy demanded to see the Croatian files he was convinced were being kept from him. He told ASIO staff that it was Labor's policy to 'bring open government to Australia'. The raid was a shambles. There were no Croatian files. Worse, the press had been tipped off about the raid by the police. Murphy's humiliation in Melbourne became Whitlam's in Canberra.

(One consequence of the botched ASIO raid was the Royal Commission on Intelligence and Security, established by Whitlam on 21 August 1974 to investigate Australia's intelligence agencies. Justice Robert Hope was named as royal commissioner. By the time Justice Hope handed down his reports in 1977, Malcolm Fraser had replaced Whitlam as prime minister.)

Murphy's reckless actions convinced the CIA that Australia's new Labor government could not be trusted with American intelligence secrets. The CIA's chief of counterintelligence staff, James Jesus Angleton, was apoplectic, calling Whitlam and Murphy 'cowboys' and describing the raid as 'one of the most extraordinary acts that one has ever seen'. In his book *Shadow Warrior: William Egan Colby and the CIA*, Randall Woods quotes Angleton saying that Washington had 'entrusted the highest secrets of counter-intelligence to the Australian services and we saw the sanctity of that information being jeopardised by a bull in a china shop'.

Three weeks after the ASIO raid, Senator John Wheeldon was voted chairman of the Joint Committee on Foreign Affairs and Defence. A classified CIA intelligence bulletin of 7 April 1973 described Wheeldon as an 'outspoken anti-US Labor senator'. Wheeldon, it said, had been a 'sharp critic of US Indochina policy and of American defense and scientific installations in Australia'.

A staunch opponent of colonialism, Wheeldon had spoken in parliament about the need for Labor to adopt a foreign policy 'more distinctively Australian and not dependent on decisions made in Washington or somewhere else'. He opposed the building of an American military base on the Indian Ocean atoll of Diego Garcia and was wary of the US Navy's plans to build an Omega navigation station in Australia, pointing out that its military value (Omega was an important navigational aid to US hunter-killer submarines, warships and military aircraft) would make it a likely target in the event of war.

But if Wheeldon was 'anti-US' he was also 'anti-Soviet'. He did not want either of the superpowers in Australia's backyard and considered the best way of keeping Moscow out was by not inviting the Americans in. 'It would be very difficult,' he told the parliament, 'to talk about getting the superpowers out of the Atlantic or the Mediterranean or the north Pacific, but it is completely feasible to talk about getting the superpowers out of the Indian Ocean'.

The CIA intelligence bulletin warned that Senator Wheeldon's election to the Joint Committee on Foreign Affairs and Defence 'could further complicate US-Australian relations' but concluded that 'it will not... endanger continued control by moderates over the Australian Labor Party and government'.

To Nixon and Kissinger, the men controlling the Australian government looked anything but 'moderate'. The president and his national security advisor were determined to 'freeze'

Whitlam 'for a few months' to make sure he would 'get the message' about the cost of his perfidy. As James Curran relates in *Unholy Fury*, Nixon banned all contact with the Australian ambassador, ensured that no New Year's message was sent to Whitlam, and sabotaged any diplomatic gestures from the State Department by ordering that all cables to Canberra had to be approved by Kissinger. Whitlam, he growled, was 'one of the peaceniks' and was putting Australia on 'a very very dangerous path'. The White House's formal response to Whitlam's letter criticising the Hanoi bombing was both cynical and mortifying: it did not bother to reply.

The ban on American ships by the Seamen's Union in Melbourne had quickly spread to Sydney and become nationwide. There were demands to punish Nixon by halting the delivery of essential supplies to the US Embassy in Canberra and calls for Australian staff to walk off the job at Woomera and Pine Gap. It looked to the Americans as if Labor ministers were inciting the boycotts. Longshoremen at US ports threatened to retaliate with bans on Australian ships.

Whitlam's refusal to speak out was widely interpreted in Australia as an abdication of leadership, but Whitlam knew better than anyone that a stoush with the unions could not end well for a Labor government. It was the ACTU president, Bob Hawke, not the prime minister, who finally negotiated an end to the bans.

The heat went out of the anti-American protests but Nixon and Kissinger were in no mood to forgive and forget. The death of former president Johnson at the end of January provided an opportunity to inflict an exquisite humiliation on the Australian prime minister. While both the Australian and New Zealand governments sent the standard letter of condolence to the White

House, Kissinger made sure that only New Zealand's Norman Kirk received a reply. The snub must have stung, but a greater ignominy was waiting for Whitlam, who was planning to visit the United States later in the year. Already it was being rumoured that Whitlam could forget any thoughts of a meeting with the president. Kissinger told a committee of the National Security Council that Nixon wanted the State Department to know that 'our relations with Australia have not improved ... Whitlam is not being invited, and if he comes anyway you can be sure he will not be received'.

At home, Whitlam put on a brave face. During a press conference at Parliament House it was put to him that there was 'a sense in Washington that Whitlam is not a Nixon type: there are far-reaching changes in Australian policy caused by Whitlam's election which ... will require some long and cosy chats between the two sides in the future'. He was then asked directly:

> [D]o you and the President see eye to eye on major issues? Are you of similar personalities or personalities so different that you will not be able to get along together?

Whitlam was canny enough to know that the prospect of 'long and cosy chats' with Nixon was some way off. He told the press, however, that he did not believe there was 'any reason why any Australian prime minister can't get along with any United States president' and that he saw 'no reason why the present prime minister can't get along with the present president'.

Whether Whitlam actually believed what he was saying or not, the CIA appeared to share his upbeat assessment of US–Australian relations. The agency was far more relaxed than the White House about doing business with Whitlam.

In a secret intelligence memorandum dated 11 January 1973 entitled 'Australia: The New Team', the CIA noted that 'much of what has been done could have been anticipated. The succession of announcements changing the policies of the outgoing Liberal-Country government grew directly out of the Labor Party's long-standing policies and election platform. The speed and firmness with which Whitlam has acted are obviously designed to serve notice on all concerned—the population, the bureaucracy, and Australia's allies—that a new team with different ideas is now in charge.' Despite the hysterics in the White House, however, it concluded that 'Australia's basic alliance with the US does not appear in jeopardy' and, furthermore, that 'the continuation of US defense and scientific activities in Australia will present no fundamental problem'.

The memorandum pointed out, however, that Whitlam's 'more independent and assertive' foreign policy would inevitably introduce 'complications and uncertainties previously unknown in the Australian-US relationship'. Significantly, the agency noted the tensions between the 'moderate' prime minister and his 'quarrelsome' colleagues and questioned whether Whitlam would be able to control 'intemperate leftist elements in his party and government', including at least one Cabinet minister with a 'long history of Communist associations'.

The CIA's judicious assessment of Whitlam's domestic predicament stood in stark contrast to Nixon's personal animus. But when the time came to choose a new ambassador to Australia, Nixon acknowledged the strategic importance of the alliance by appointing Marshall Green, a high-ranking career diplomat, rather than the customary political appointee that Australia had become used to. In *Unholy Fury*, James Curran describes Green being sent on his way with 'a string of presidential "expletives" about Whitlam', culminating in the words 'Marshall, I can't stand that [****]'.

In the first six months of 1973 Nixon would entertain King Hussein of Jordan, Singapore's prime minister, Lee Kuan Yew, and Ethiopia's Emperor Haile Selassie, but Whitlam continued to be frozen out.

The latest snub reminded the Australian press of the embarrassment Whitlam had suffered a year earlier when, as opposition leader, he had spent two and a half days in Washington waiting for a meeting with Nixon that never took place. On the earlier visit Nixon had made time to see the West German opposition leader but not Whitlam; while Kissinger, whom Whitlam had also hoped to meet, preferred attending a black-tie function with Cristina Ford, the Italian socialite wife of the Ford Motor Company boss, to meeting the Australian Labor leader.

Sydney's *Sun-Herald* newspaper cast doubts on Whitlam's later denial that he felt 'slighted' by President Nixon 'ducking out of a meeting with him' in 1972. Acording to the *Sun-Herald*, Whitlam had been 'piqued' and made this clear at a dinner he had with senior business executives in New York. Its New York correspondent, Derryn Hinch, claimed that Whitlam 'was visibly and volubly angry at journalists when questioned over why the President hadn't seen him'.

An official announcement in March 1973 that Whitlam would 'almost certainly' visit Washington in August en route to Ottawa for the Commonwealth prime ministers' conference prompted further speculation over whether he would actually meet the president. By taking on the foreign affairs portfolio, Whitlam had put himself 'in the unfortunate position of obvious supplicant for a White House invitation', the *Sydney Morning Herald* commented, while Nixon 'as anticipated ... was non-committal'. The president, it said, had 'simply not decided whether he wants to see Mr Whitlam, or is prepared to put himself out to receive him'. There was even talk that the next ANZUS meeting might have to be switched to New Zealand to save Whitlam from further humiliation by Nixon.

Six months after becoming prime minister Whitlam stood up in the House of Representatives and praised Nixon for his 'pivotal role in ushering in a new and saner phase in our relations with China', for his achievements in the Strategic Arms Limitation Talks, and for pulling the last US troops out of Vietnam. As well as repeating his promise to renegotiate the deal for the US Navy communications station at North West Cape, he reiterated the value of the US bases at Pine Gap and Nurrungar and confirmed Australia's commitment to the ANZUS Treaty. The speech was a clear attempt by Whitlam to mend fences with the Nixon White House, but it went unacknowledged.

Chapter 15
NSSM 204

In the first week of June 1973, Lance Barnard attempted to resolve the impasse by calling on the new US ambassador. Marshall Green described the visit in a cable to Washington. The cable began:

> Acting Prime Minister and Defense Minister Lance Barnard called at residence today. He explained that this unusual confidential visit in advance of my credentials presentation was his personal initiative. Neither Prime Minister nor other members of Government were aware of his actions, he said.

The purpose of Barnard's intervention was to enlist the ambassador's help in getting Whitlam through the door of the White House. The question of whether or not Nixon would receive the prime minister in Washington had become 'an emotional issue in Australia', Barnard said. He understood that Nixon had taken grave offence at Australian comments about the Hanoi bombings and freely admitted that 'the statements made were indeed offensive'. However, he said, Nixon was perhaps not aware that he and Whitlam 'had called the three offending ministers before them for a stern dressing down and warning'. As for Whitlam's own remarks, Barnard told the ambassador that in the heat of debate the prime minister 'sometimes made comments he later regretted'.

While Barnard expressed his concern for the future of the US–Australian alliance in general, his appeal to Green rested on the issue of the defence bases. According to Barnard, as long as Nixon left him out in the cold, Whitlam would have a tough job defending the bases against attacks by the Labor left. Although he and Whitlam had no doubts about the 'essentiality' of Pine Gap and the other defence facilities, Barnard warned Green that they 'would again be under attack' at the ALP conference in July. If at the conference Whitlam was still unable to state that he would meet with the president, this fact 'would be utilized . . . to cast doubt upon importance US itself attached to alliance'.

The ambassador told Barnard that he was 'honored by his visit' and 'impressed by the evident sincerity and honesty of his remarks'. In his cable to Washington, Green described the meeting as a 'moving and, I believe, entirely candid discussion'. While Barnard 'did not attempt to blackmail us by claiming the future of our installations was at stake', Whitlam's deputy made it clear that left-wing hostility to the US–Australia alliance always crystalised around the bases. Green duly passed on the message: the continuing freeze on Whitlam was being viewed in Australia as an 'insult'. Pine Gap was safe, at least for now, but the White House meeting had to happen.

The behind-the-scenes lobbying finally paid off. On 17 June 1973 Nixon gave in to the urgings of his ambassador and announced that the Australian prime minister would be received at the White House on 30 July.

Just before he was admitted to the Oval Office, Whitlam told Kissinger that 'in Australia ... the new prime minister still must get his legitimacy within the first few months by gaining US accolades'. The 'accolades', when they arrived, were perhaps less effusive than Whitlam had hoped. At the long-awaited meeting, which was originally scheduled for an hour but (according to the official memorandum of conversation) ran twenty minutes over, Nixon said he had 'never met an Aussie I didn't like' and told Marshall Green that he had found Whitlam 'quite a guy'. Kissinger was less impressed, later telling Singapore's Lee Kuan Yew that Whitlam was 'not a heavyweight'.

But Whitlam had done what he set out to do. He had shed the aura of sycophancy inherited from his conservative predecessors and put the US–Australian relationship on the path towards a 'new maturity'. The message he wanted to leave with the White House was that as long as he was in charge, the alliance would remain sound and the bases would be protected.

The Americans, however, were unconvinced. Kissinger did not trust Whitlam, and nor did the CIA chief, James Schlesinger. 'Whitlam's a bastard,' Schlesinger declared at a Pentagon breakfast meeting in September 1973 (by then he had been made Secretary of Defense). Kissinger replied, 'I agree.' Schlesinger accused Whitlam of 'playing games'.

In January 1974 Lance Barnard made a courtesy call on Vice President Ford at his office in Washington's magnificent late-19th century Old Executive Office Building. State Department briefing notes prepared for the vice president before the meeting state that in foreign policy, the Whitlam government 'has tended to be moralistic and activist, although somewhat doctrinaire'.

On the subject of US bases in Australia, Ford was advised that 'the importance of our defense installations in Australia is accepted and defended by Whitlam, and the "American alliance" remains the cornerstone of Australia's defense and foreign policies. When they offer criticism, they do so (in their view) as a privileged old friend and ally—from "within the family".'

Under 'talking points', the vice president was encouraged to 'initiate a discussion of bilateral, regional, and global issues' with Barnard and to note that 'although the US and Australia may differ from time to time on specific issues, it is important that we keep in close and intimate touch, at all levels of government, in order to have a full understanding and appreciation of our respective viewpoints and concerns'.

Barnard—described in the vice president's briefing notes as an 'intelligent, open-minded man' and 'probably the best friend we have in the current Labor Government'—was accompanied to the meeting by Australia's ambassador, James Plimsoll, and Arthur Tange. Marshall Green was also present.

With Barnard keen to show that Labor could work with the United States, and the Americans anxious not to derail negotiations over North West Cape, the meeting went smoothly. An official memorandum of the conversation notes that Barnard invited the vice president to visit Australia and told him that Australians were 'very grateful and will always remember the contribution of the United States to Australia's security during World War II'. Barnard also assured Ford that he 'prizes good relations between our two countries very highly and is deeply aware of the importance of the ANZUS Treaty'. (Unfortunately, the memorandum, stamped 'confidential' and carrying the crest of the office of the vice president, mistakenly refers to Barnard throughout as the 'prime minister'.)

At home, the 'games' over the US bases continued. On 3 April 1974 Whitlam was asked in the parliament about a proposal to build a 'joint Australian-Soviet station'. An answer to the question had been drafted for Whitlam by staff in the foreign affairs department. Noting that such a plan would have obvious 'implications for various United States installations in Australia', Whitlam explained that the plan was 'currently under study in appropriate departments'. He then veered away from his prepared reply with an incendiary statement:

> The Australian Government takes the attitude that there should not be foreign military bases, stations, installations in Australia. We honour agreements covering existing stations. We do not favour the extension or prolongation of any of those existing ones. The agreements stand, but there will not be extensions or proliferations.

That the ruling out of 'extensions' to existing agreements was an off-the-cuff remark, a departure from the script, made it more, not less, ominous to the Americans, because Whitlam was in charge and his words appeared to have the weight of personal conviction.

In May 1974 Whitlam led Labor to another federal election victory. While the Coalition gained three seats, Labor retained a narrow majority. The left-winger Jim Cairns—who made no secret of his desire to get rid of all US military installations in Australia—was elected deputy prime minister.

The elevation of Cairns alarmed US security chiefs already struggling to make sense of Whitlam's vacillations. At a conference on relations with Australia, chaired by acting secretary of state Joe Sisco just four days after Cairns's election as deputy prime minister, 'considerable concern' was voiced over the ousting of Lance Barnard. While nobody had been able to determine whether Cairns was a communist, the conference heard that he had 'sided with Hanoi and Peking in many issues' and had 'said unequivocally that all foreign military installations should be removed from the country'. The conference was told that consideration was being given to the 'possible relocation of two of the projects there' (probably Pine Gap and Woomera) and that this 'would require a tremendous lead time, a lot more money'.

A week later the initial panic over Cairns's elevation had subsided, due in part to Arthur Tange's reassurance that, even as deputy prime minister, Cairns would not necessarily be given access to sensitive intelligence. At a meeting on 21 June, deputy assistant secretary of state Art Hummel argued that the granting or withholding of security clearance for Cairns was Whitlam's problem and the United States should not become involved. '[W]e don't tear down the relationships we have by arbitrary action on our part, so that we don't engage in spooky fiddling with the situation, in which we might get caught,' Hummel said. Such interference was 'logical to consider, but we think not logical to carry out'.

Cairns's election as deputy prime minister nevertheless signalled a lurch to the left that threatened to have severe implications for the alliance.

On 1 July 1974 the White House commissioned a top-secret study of US relations with Australia 'in the light of recent changes in the Labor Government'. National Security Study Memorandum (NSSM) 204 would be a multi-agency study drawing on the collective wisdom of the CIA, the state and defence departments, the Joint Chiefs of Staff, the National Security Agency, the

National Security Council and the Defense Intelligence Agency. Henry Kissinger's tasking memorandum listed key issues to be examined. These included:

- The implications of changes in the Australian government for future relations between Australia and the US.
- The prospects for keeping US defense installations in Australia, and the policy options for trying to prolong their existence there.
- The alternatives for relocating essential existing US security functions outside of Australia, and the impact on our alliance relationship of doing so.
- The prospects for locating additional US defense installations in Australia, and the policy options for trying to do so.
- The risks involved in the continued sharing of intelligence with Australia, the alternative means for reducing these risks, and the impact on our alliance with Australia of curtailing or ending such intelligence sharing.

That the terms of reference revolved around America's military installations in Australia was, as James Curran points out, 'a telling indication of the supreme importance Washington attached to the facilities. For the United States, the facilities came first: those facilities defined the alliance.'

NSSM 204 was a top-to-bottom analysis of the deteriorating US–Australian relationship. While it found that Australians were 'increasingly sensitive to implications of US domination', it concluded that they were 'not anti-American'. Most Australians, it said, 'want relations with the United States and with Americans to remain close and friendly'. The Labor government was expected to cooperate with the United States on security matters and to keep existing facilities 'at least until the expiration of the present agreements', while pushing for a greater say in how the facilities operated. The review noted that while Whitlam 'would prefer to see the eventual departure of the US installations', he 'agrees that they should remain for the time being'.

What really frightened Washington was potential subversion of the Labor Party by the left wing led by Cairns, a figure characterised in NSSM 204 as 'Marxist by persuasion'. The review warned that 'one or more of these men might leak security or defense information in order to embarrass both Whitlam and the Australian-American alliance'. Using their new positions of power, the 'leftists' were expected to 'criticize the Australian-American alliance' and call for 'a withdrawal of all US security facilities from Australia'. While it was unlikely they would succeed, their efforts were expected to 'disturb the US-Australian relationship' and 'undermine general Australian acceptance of the American presence'. (As it turned out, Cairns did not keep his job long enough to constitute a serious threat to the alliance: in July 1975 he was sacked from the ministry by Whitlam for misleading parliament. The new deputy prime minister was Frank Crean.)

Pine Gap (described as 'the only ground station for a classified military satellite') was the subject of a detailed—and still classified—annex to the main report, but NSSM 204 warned that it was 'less than certain that the GOA [Government of Australia] will continue to provide this installation a favorable, protected environment, or that it will not exercise its option to terminate the existing agreement upon one year's notice anytime after December 1975'.

A graphic summary of 'defense relationships and benefits'

revealed that the two sides did not benefit equally from the 'joint' bases. Pine Gap was assessed as delivering 'exceedingly high' benefit to the United States, but only 'high' benefit to Australia. Woomera was judged to have 'very high' benefit to the US, but only 'high' benefit to Australia. (By contrast, the exchange of military technology was assessed as only of 'medium' benefit to the US but of 'very high' benefit to Australia, with 'unique information involved'.)

NSSM 204 laid out a series of three 'basic policy options . . . directed towards maintaining our basic alliance relationship with Australia'. Options 2 and 3 were variations on 'business-as-usual', with some carrots (such as a possible visit by the vice president) thrown in. Option 1 was much more aggressive:

> Begin immediately to attenuate certain ties in the US-Australian alliance relationship, on the assumption that this will induce the Whitlam Government to reverse those major elements of its foreign policy which are inimical to US interests.

This policy could be achieved by 'one or more' of the following steps:

- Reduce as soon as possible the flow of the most sensitive intelligence to the GOA.
- Reduce the intimacy of official relations with the Australians in those fields of activity that are not vital to the US.
- Undertake immediately some reduction in joint US–Australian military exercises.
- React vigorously to GOA statements and policy initiatives that seem to contradict the 'ANZUS relationship'.

The 'pros' of the policy were that it would make 'crystal clear' US dissatisfaction with Whitlam's foreign policy and might damage Whitlam's standing with voters, 'setting the stage for an Opposition victory'. The 'cons' were more numerous and included the risk that new US pressures would 'undercut our friends within Australia'; that they might 'provoke a general nationalistic Australian reaction' against the US; that they could prompt a 'downward spiral' in the US–Australian relationship; and that they could 'interfere' with efforts to alter Pine Gap arrangements in ways most suitable to US interests and might result in 'early loss of that facility and/or others'.

As well as the 'basic policy options', NSSM 204 canvassed a range of policy options for the installations themselves. Again, some of the options were variations on the status quo. Others involved attempting to 'obtain GOA agreement to extend the present arrangement until 1978' or, more radically, notifying Canberra of an intention to withdraw after December 1976 (i.e. twelve months after the expiry of the existing lease).

The benefits of abandoning Pine Gap, either after 1976 or after 1978, would be to 'eliminate a potentially major friction in US-Australian relations' as well as 'hopefully reduce left-wing pressures against other US defense installations'. On the minus side, a US withdrawal from Pine Gap would have the appearance of a 'victory for the ALP left wing' and could have a 'negative affect on the entire ANZUS relationship' by suggesting the US was losing interest in the alliance.

The Watergate cover-up forced Nixon to resign before he could study the review. Vice-President Gerald Ford took over the presidency on 9 August 1974, with Kissinger staying on as his secretary of state.

In Washington, NSSM 204 provoked intense debate and

became the subject of futher reviews and analytical studies. On 24 August it was discussed at a meeting of the Senior Review Group (SRG), a high-level body established by Nixon to review papers before their submission to the National Security Council or the president. The SRG's members included the CIA director and the chairman of the Joint Chiefs of staff. The purpose of the meeting, outlined in a top-secret memorandum to Kissinger, was to discuss 'whether we should modify our policy toward Australia because of Canberra's continuing turn to the left and whether we should plan to shift some of our installations elsewhere'.

The Kissinger memorandum outlines the 'problems' with the Whitlam government and notes, 'if we are confident that Australia's independent course will not go too far, we would keep our installations. If not, we would move them. The former is more risky; the latter more costly.'

According to the memorandum, Labor's shift to the left had prompted the Department of Defense to 'study when and where to relocate its installations', drawing a desperate reaction from Australian officials in Washington who 'pleaded that we should not overreact'.

While the Department of Defense was now believed to be 'less concerned than it was', the State Department 'still believes that we are better off if we do not rock the boat in these tricky waters; it believes all will be well if we continue to act as though our basic relationship remains as before and do not assume the worst'.

While the CIA had 'no clear [policy] position' on the US–Australian alliance as a whole, it was said to be pessimistic about the chances of holding on to Pine Gap, hoping merely that 'there might be a chance of leaving it in Australia into the 1980s'.

The hard-nosed US ambassador, Marshall Green, was prepared to write off Pine Gap altogether in return for defence cooperation in other areas. The Kissinger memorandum expressed support for Green's view that defence cooperation with Australia could be expanded 'if we agree now to remove Pine Gap in 1978'.

Even holding on to Pine Gap until 1978 was not considered a certainty, although there was a feeling within the National Security Council that it might be possible to 'lower the political risk of keeping the Pine Gap operation in Australia until 1978 by making the operation more genuinely a joint US–Australian venture, as Whitlam wants'.

Whitlam's unpredictable personal comments continued to keep the Americans on edge, but in late August it was felt the 'immediate threat to our installations has eased somewhat'. By the end of 1974 Marshall Green was rejoicing at the Labor leader's ability to go five months without uttering any 'gratuitous remarks'. James Curran quotes a conversation in Washington between Marshall Green and James Schlesinger, now President Ford's secretary of defense, in which Green accepted that on the defence installations, Whitlam was 'on our side'. Kissinger, who once considered Whitlam a 'bastard', now conceded he was 'mellowing'.

Chapter 16
Never again

By the second half of 1975 the Labor government was unravelling, its credibility shot by a series of ministerial sackings and resignations; fecklessness in the face of rising inflation and a bizarre attempt to bolster Australia's flagging economy with a $4 billion loan in Arab 'petrodollars' that was never delivered. Looking for ways to attack a resurgent opposition, Whitlam rounded on an old enemy.

In October Labor Party staffers reportedly received disturbing information about an American named Richard Lee Stallings. Stallings, a retired CIA officer, had arrived in Australia in September 1966 and played a crucial role in negotiating the establishment of the Joint Defence Space Research Facility at Pine Gap. A year later he would become its first chief of facility.

According to scuttlebutt that had reached the ears of Whitlam staffers, Stallings had a fondness for talking about his CIA connections and his work at Pine Gap. Such talk had the potential to be highly embarrassing to both governments, since Pine Gap had been publicly identified as a joint project with the US Department of Defense, not the CIA. To opponents of Pine Gap, the involvement of the CIA suggested that the station was involved in more nefarious activities than had been admitted. Arthur Tange noted in his memoir that Washington 'took precautions to avoid speculation when any senior official in their intelligence community visited the Facilities' but was 'open' about the visits of defence officials.

When Whitlam asked the Department of Foreign Affairs for the names of covert CIA officers working in Australia, Stallings's name did not appear on the list. It did appear, however, on a list of CIA agents supplied by the Department of Defence.

Incensed by the double deception, Whitlam prepared to lash out in a way that, as the CIA's East Asia bureau chief later put it, threatened to 'blow the lid off' Pine Gap and every other US intelligence facility in Australia.

On 2 November, just nine days before he was sacked by the governor-general, Whitlam visited Port Augusta in South Australia and told the audience at a Labor Party rally that the leader of the National Country Party, Doug Anthony, had received 'CIA money'. He also accused other opposition MPs of being 'subsidised by the CIA'.

In the parliament, Anthony furiously denied Whitlam's claim, as well as unsubstantiated allegations in the press that the CIA had funnelled money into Australia with the aim of 'influencing Australia's domestic political situation with a view to protecting America's interests'. After Brian Toohey, in the *Australian Financial Review*, revealed that Anthony had once rented his Canberra home to Stallings, and named Stallings as the former CIA chief at Pine Gap, Anthony accused Whitlam of a 'cowardly and gutless' attack. Anthony claimed to know nothing of Stallings's CIA affiliation or his role at Pine Gap;

as far as he knew, Stallings worked for the US Department of Defense. Mr and Mrs Stallings turned out to be 'very charming people' and the two families occasionally exchanged Christmas cards. Whitlam responded by calling Anthony 'politically gullible'.

After a US State Department official was quoted denying that Stallings had worked for any US intelligence agency, Anthony challenged the prime minister to provide evidence to the parliament that Stallings was a CIA officer.

Sir Arthur Tange, head of the defence department, watched in horror as the country's defence secrets were heedlessly bandied about by feuding MPs. Tange had grown used to Australian journalists and academics 'speculating' about intelligence matters, but having their speculations confirmed in the national parliament was far more serious.

Desperate to keep the Soviets from finding out (if they did not already know) that Pine Gap was run by the CIA, not the US defense department, Tange warned Whitlam (according to journalist Paul Kelly) that providing such evidence would anger the Americans to the point of 'jeopardising the alliance'. While imploring Whitlam to keep to the line that Stallings worked for the Pentagon, not the CIA, Tange tried to exert pressure through Whitlam's staff, telling one aide that the stoush with the Americans over Stallings represented 'the gravest risk to the nation's security there has ever been'.

On 10 November Whitlam was handed a cable sent by the ASIO liaison officer in Washington to the acting head of ASIO in Melbourne. The cable carried a warning from Ted Shackley, head of the CIA's East Asia division, about the damage that would be caused to US–Australian relations if Whitlam carried out his threat to reveal Stallings's role with the CIA. According to Shackley, the agency felt that 'if this problem cannot be solved they do not see how our mutually beneficial relationships are going to continue'. In particular, the CIA was worried about the effect of Whitlam's disclosures on intelligence operations at Pine Gap.

The cable soon landed on Tange's desk. Dismissing Shackley as a 'ham-fisted American intelligence official', Tange decided that 'this man's threats were not a matter for concern' and that Shackley's superiors in Washington could be trusted to 'hose him down'.

Hosing down Whitlam was another matter. He was scheduled to give his answer to Anthony's question on the afternoon of 11 November. Before he got the chance, the governor-general, Sir John Kerr, sacked him.

Had the row over Stallings, Pine Gap and the CIA played any part in the dismissal? Tange would later be accused of having arranged for Dr John Farrands, chief defence scientist and the only bureaucrat Tange trusted with the secrets of Pine Gap, to brief Kerr on the CIA's concerns about the security of its intelligence facilities in Australia.

Tange, Kerr and Farrands all denied it, and in the years to come journalists would be warned off the story of an alleged conspiracy between Kerr, Tange and the CIA with threats of legal action. Whitlam scoffed at the idea of Kerr, who had links to Australian intelligence during the Second World War, sacking him at the behest of the CIA, although his wife, Margaret, was less sure.

In his book *The Whitlam Government 1972–1975*, Whitlam recalled a meeting at Sydney airport with the State Department's assistant secretary for Asia and the Pacific, Warren Christopher, in 1977. According to Whitlam's account, Christopher had been asked by President Jimmy Carter to assure him that 'the US administration would never again interfere in the domestic political process of Australia'.

No solid evidence of a CIA conspiracy has ever been found, but the theory refuses to die.

Four weeks after Whitlam's removal, Malcolm Fraser led the Coalition to a huge victory in the federal election. With Sir Arthur Tange staying on as his secretary for defence, Fraser was determined that the gates to Pine Gap remain closed. Justice Hope, appointed by Whitlam in 1974 to lead a royal commission into Australia's intelligence community, had planned to visit Pine Gap in the course of his investigation but was denied entry. The CIA was still in charge, and for years it was believed that the Americans were behind the decision to keep him out. However, declassified documents released by the National Archives in 2008—including some marked 'TOP SECRET AUSTEO [Australian Eyes Only]'—reveal that it was the Fraser government, not the CIA, that barred Hope from Pine Gap.

After the royal commission delivered its reports to the Fraser government in 1977, the commission's secretary, George Brownbill, wrote a memo to Sir Geoffrey Yeend, then head of the Department of Prime Minister and Cabinet, noting that Justice Hope had asked Tange for permission to visit Pine Gap but that '[n]o reply was ever received'. However, Brownbill wrote, '[w]e knew . . . that the request had created a great stir in Defence'. Far from being rebuffed by the Americans, 'we had many conversations with US officials about [Pine Gap] and found them far readier to provide information than the Australian department'. (Tange remembered things rather differently, declaring in his memoir that '[w]e took it as a duty to answer the Judge's questions'.) More than 30 years later,

Brownbill repeated his assertion that Tange had resisted giving information about Pine Gap to the Hope commission. 'To do so, it seemed to Tange, would put the Western alliance in jeopardy,' Brownbill told the *Canberra Times*.

Sir Arthur Tange would stay on as head of defence for two more years after the release of the Hope report, but in that time not even he could keep the secrets of Pine Gap safe.

Chapter 17
Drop that hamburger!

In July 1976 Malcolm Fraser and his foreign affairs minister, Andrew Peacock, were sitting in the White House with President Ford and Kissinger. The future of Pine Gap was one of the first items on the agenda.

A declassified memorandum of the conversation suggests a convivial meeting, with Fraser quick to distance himself from his predecessor. Australia's new prime minister told Ford that he had been 'particularly' keen to visit in order to 'wash away a few vestiges of difference between our countries'. Ford responded, 'I very much appreciate the change.'

A sign of improved US–Australian relations was that US nuclear-powered warships, which had been barred from Australian ports since 1971, were again allowed to visit. After cutbacks under Labor, the Fraser government was also committed to lifting defence spending, a policy that was always calculated to win favour in Washington.

Noting that the renewal of the Pine Gap lease had previously been considered on an annual basis, Fraser proposed that in future renewal should be done on a ten-year basis 'if you would find that useful'.

'That would be very helpful,' Ford told him.

What neither Ford nor Fraser knew as they exchanged diplomatic niceties in the White House was that the future of Pine Gap—and of the whole US–Australian intelligence program—were about to be put in jeopardy by events taking place 2500 kilometres away in Mexico City.

On 5 January 1977, six months after the White House meeting between the US president and the prime minister of Australia, Andrew Daulton Lee, the adopted son of a wealthy Californian pathologist, went to an intersection near the Miguel Aleman Freeway in Mexico City and marked a lamp post with a letter 'x' made from adhesive tape. It was Lee's regular sign that he had another delivery for a man he knew as 'John' but who was really a KGB agent named Boris Grishin, who masqueraded as a science attaché at the Soviet Embassy.

After leaving his marker, the usual routine was for Lee to meet Grishin that evening at a pizza restaurant. When Grishin failed to show up, Lee followed a pre-arranged procedure and returned to the restaurant the next morning. Again, there was no sign of Grishin. Defying strict instructions never to visit the embassy, Lee then did exactly that.

The Soviet Embassy in Mexico City had long been a magnet for would-be spies and other disaffected souls wanting to make contact with Russians. Lee Harvey Oswald visited both the Soviet and Cuban embassies in Mexico City in the weeks before he assassinated President Kennedy. Both the CIA and the Mexican police kept the Soviet Embassy under close watch. It was the Mexicans who noticed Lee toss a pink envelope into the embassy grounds and saw it being collected by an embassy official.

The Mexican police promptly arrested Lee, who tried to hide the joint he was carrying. Refusing a bribe of $500 to let him go, the police searched Lee and found microfilm negatives of documents marked 'Top Secret'. They did not believe his explanation that he worked in advertising and was using the documents in advertisements for the General Electric Company.

At around midday on 6 January 1977 a vice-consul at the US Embassy was summoned to police headquarters. Such requests were common; the official spent much of his time dealing with Americans picked up by the Mexican police for drug offences. When the vice-consul arrived he was shown photographs made from the negatives Lee had been carrying when he was arrested. He immediately called a colleague at the embassy who worked for the CIA.

Ten days later, Lee's childhood friend Christopher John Boyce was arrested by FBI agents on suspicion of being a Soviet spy. Boyce, the son of an FBI agent, had a passion for falconry. He was captured near the University of California campus at Riverside after hunting all day with his falcon.

Lee had been in trouble before. He had been arrested a few times for possessing and selling drugs and had served several months in jail for selling cocaine to undercover Los Angeles police. Boyce was smart, but something of a drifter. In 1974, after dropping out of college for the third time, he took a job at TRW Inc. of Redondo Beach, the aerospace firm that designed and built spy satellites for the CIA, where a former FBI colleague of his father's was the director of security.

In his book *The Falcon and the Snowman*, Robert Lindsey describes Boyce's introduction to the work done inside Building M-4, a place off limits to most TRW employees. Boyce was briefed by a supervisor named Larry Rogers. According to Lindsey, Boyce was high on amphetamines.

> Rogers ordered Chris never to discuss with anyone the briefing he was about to give; never to divulge to anyone the existence of the projects of which he was about to hear, or discuss with anyone the kind of work he was to do; and never to mention to anyone not cleared for the projects their code names or the fact that they were clandestine operations of the Central Intelligence Agency.
>
> In fact, he ordered, never mention to anyone—his family, his girlfriends, any outsiders—that the CIA had any relationship whatsoever with TRW, or that his salary was being paid by the CIA under a contract with TRW.

Nominally hired as a 'communications clerk', Boyce soon found himself in charge of the 'black vault' or code room. Only eight people, Lindsey writes, were allowed to enter the black vault. All had to clear the highest level of CIA vetting. Before entering the code room, they had to pass through three checkpoints. CCTV cameras kept them under constant surveillance. Boyce's security went even higher than 'top-secret'; he was given 'crypto clearance', allowing him to handle coded material. At the age of just 21, the college dropout found himself at the heart of one of the US government's most highly classified intelligence programs. Satellites built by TRW could photograph Chinese missile bases and Soviet submarine pens using cameras so sensitive that they could capture a man walking his dog from a hundred miles up in space.

Lindsey's book describes a space-based intelligence program consisting of three parts, all compartmentalised so that staff working in one section would not have access to secrets involving other sections.

> One component was assigned to build and operate the satellites; the second was responsible for collecting and initially processing the data... sent to earth by the satellites; the third was a massive, on-the-ground program to analyze the data... There were more than a dozen different types of satellites, each with its own project code name, mission and method of operating; each system might have three or more different satellites... each sending back 'product' for analysis.

One of the systems Boyce learnt about was Project Rhyolite, controlled from a ground station in central Australia.

Boyce's job in the code room at TRW involved sending encrypted messages about the satellites—as well as other sensitive military information—to the CIA's headquarters at Langley and other destinations around the world.

Des Ball estimated that 25 to 30 messages were sent from Pine Gap to TRW each day, about half of which were to be relayed to Langley, and around the same number were sent from TRW to Pine Gap. A significant part of the traffic from Pine Gap to TRW involved data about the performance of the Rhyolite satellites controlled from Australia. According to Ball:

> Analysis of the telemetry from the SIGINT satellites enables TRW engineers to monitor the 'health' of the satellites and their hydrazine fuel consumption, and hence to estimate their expected operational longevity... Regular analysis of the characteristics of intercepted signals enables TRW

engineers to suggest changes in satellite control techniques and software programs to enhance the intercept capabilities of the SIGINT satellites; this sometimes requires the involvement of these engineers in intercept operations in as close to real-time as possible.

The messages sent to CIA headquarters were largely about administrative matters. In *The Falcon and the Snowman*, Lindsey mentions telex messages about industrial action in Australia that could 'disrupt movement of equipment and personnel to Alice Springs' and about CIA efforts to prevent strikes or find ways to minimise their impact on Pine Gap.

As well as relaying encrypted messages, Boyce was cleared to handle the codes themselves, including the unique daily computer cards that controlled each day's encryption settings. It was these that piqued the Russians' interest when Andrew Daulton Lee wandered into the Soviet Embassy in Mexico City one day in 1975 with a story about his friend who worked in the code room at TRW. Lee was given money to buy a miniature camera and for a year and a half Boyce used it to photograph secret documents inside the black vault. He even smuggled documents out of the vault in potted plants. At his trial Boyce would say, 'Security at TRW was a joke.'

As well as the daily code settings, the Soviets were eager for information about the satellites used to detect missile firings—in particular a TRW satellite (probably a Rhyolite downloading data to Pine Gap) that photographed Soviet missile sites 'two or three times a day'.

Boyce was told about the executive agreement between the US and Australian governments to share intelligence obtained from the satellites and was outraged, according to Lindsey, to discover that the CIA was reneging on the deal. The United States, his briefing officer told him, 'was not living up to the agreement: certain information was not being passed to Australia ... TRW was designing a new, larger satellite with a new array of sensors; the Australians ... were never told about it; anytime [Boyce] sent messages that would reach Australia, he must delete any reference to the new satellite'. (The 'new' satellite was the Argus, launched in June 1975.)

Boyce left TRW just weeks before his friend Lee made his fateful attempt to extract a final payday from the Soviets. Four months after his arrest at Riverside, California, Boyce was put on trial for espionage in the Los Angeles federal court. Defying attempts by government lawyers to muzzle him, Boyce declared that he was 'absolutely appalled' by the 'dishonest, deceitful' practices carried out against 'one of our closest allies' and that this had motivated him to become a spy. Boyce gave evidence on oath that in his briefings at TRW he had learnt that 'things were going on from the CIA at TRW whereby we were breaking agreements with one of our allies, Australia' and that TRW employees practised 'day-to-day deception in our transmissions to the Australians'. Before Boyce could give any details, the judge shut him down.

It did not take long for Australian parliamentarians to get their hands on a transcript of Boyce's evidence, but the transcript revealed nothing more than had already been published in newspaper reports of the trial. All details of the CIA's alleged 'deception' were suppressed. Outside the court, however, Victor Marchetti, the former senior CIA officer who had helped draft the secret treaty covering the use of Pine Gap, corroborated some of Boyce's most damaging revelations, confirming to the ABC's *Four Corners* program that the existence of the Argus satellite and its enhanced capabilities had been concealed from the Australian government.

Marchetti also rekindled the still-smouldering controversy over alleged CIA involvement in Australian politics. In an interview with the Sydney *Sun*, Marchetti alleged that for at least a decade CIA money had been used to undermine the Australian Labor Party and to help keep the Coalition in power. According to Marchetti there were six or eight 'upfront' CIA operatives in Canberra and another 20 to 30 'deep cover' or clandestine operatives working elsewhere in Australia.

The comments by Boyce and Marchetti were enough to set the hares racing in parliament. On 26 May Labor's Senator Jim Cavanagh asked the government a series of questions about the CIA and Pine Gap.

> (1) Is it a fact that the inner bunker system and monitoring equipment of the defence system at Pine Gap can be reached only by United States Defense Department personnel having the highest security clearance. If so, how does the Australian Government ensure that projects carried out within this area comply with the agreements made between Australia and the United States for the establishment of the Base. (2) Could the United States be using the defence base at Pine Gap for projects without the Australian Government's knowledge.
>
> (1) Can the Minister confirm reports that the Central Intelligence Agency has set up a secret 'super bug' at Pine Gap which can monitor any telephone or telex message into or out of Australia.

The standard response from governments on both sides was to refuse either to confirm or deny 'speculation or assertion' about Pine Gap. A week later, however, the defence minister, Jim Killen, made a statement outlining, for the first time, how the spy station was run.

Killen said that both Australians and Americans working at Pine Gap had 'equal rights of access to all parts of the facility and to its activities and all results of the research. This right of access excludes only the two national communications rooms of the operating partners, to which, in accordance with standard practice at all joint installations, access is restricted to preserve national cipher security. National privacy in this respect, however, does not, and cannot, extend to denial of knowledge of the programs and activities of the facility.' The senior Australian on site, Killen said, 'works closely with his United States counterpart, participating daily in the decision-making concerned with program activities ... He has detailed knowledge and experience of the capability and activities of the facility and, of course, complete access at any time to all areas and operations ... There is no way in which a systematic deception or activities detrimental to Australian interests could go undetected, even were it assumed that it was feasible and advantageous ever to attempt them. Programs are actively monitored, both at the facility and by the Department of Defence in Canberra, to ensure compatibility with Australian national policy and interests. Australia has the right to intervene if it has doubts or objections regarding any activity.'

Killen's statement directly contradicted the evidence given by Christopher Boyce on oath at his trial, as well as Victor Marchetti's comments to the Australian media. If he was speaking from personal knowledge then it appeared Killen knew more about the facility than at least two prime ministers. By his own account, John Gorton had been happy to be kept in the dark about Pine Gap during his time as prime minister. While Gorton's successor, William McMahon, believed he knew what went on at Pine Gap and Nurrungar during the time he was prime minister, he admitted in May 1977 that he was no longer sure. 'I have increasing doubts that the Australian Government knows the entire truth,' McMahon said.

During his three years as prime minister, Gough Whitlam claimed to know everything that happened at Pine Gap. Asked at a press conference in October 1974 whether he intended to allow the United States to keep its intelligence-gathering and communications facilities in Australia, Whitlam replied, 'My Government knows what the United States is doing in Australia, and we know that nothing the United States does in Australia will be done except with our full knowledge and our full concurrence.'

He repeated this mantra a few months later in Australia, telling an ABC news reporter, 'We know what goes on there and we can stop it if we want to. There is no derogation of Australian sovereignty.'

But in the aftermath of what Whitlam liked to call the 'coup' of 11 November 1975, doubts set in. The revelations of CIA duplicity over Pine Gap that emerged from Boyce's trial appeared to confirm that Whitlam and his government had been lied to. 'One does not condone Boyce's activities, but one cannot ignore his evidence,' Whitlam told the parliament on 4 May 1977. Noting that US government lawyers at Boyce's trial had made 'strenuous, and successful' efforts to suppress his evidence, Whitlam described Boyce as 'a man in the know, a man with inside knowledge and personal experience of espionage activities ... He had access to information about the activities of the Central Intelligence Agency. How far can we afford to ignore his evidence or to rubbish his credibility?'

As a result of Boyce's disclosures, Whitlam was forced to concede that important facts about Pine Gap had been kept from him when he was prime minister. He acknowledged that information collected at Pine Gap had been given to TRW Inc. in California without the knowledge or agreement of the Australian government; that the first head of facility at Pine Gap was not an employee of the US defense department, as he had been led to believe, but a senior CIA officer; and that Pine Gap had, from the beginning, been a CIA operation.

The Fraser government continued to 'rubbish' Boyce's credibility, but Boyce's claims that the CIA was running Pine Gap in defiance of the sharing agreement echoed Leonce Kealy's comments to the *Canberra Times* in 1977 that 'the Americans run that place' and that Pine Gap was 'not meant to be a place where Australians can feel comfortable'. If Boyce's evidence was correct, little had changed in the power balance since Kealy arrived in 1970 to find Pine Gap full of Australian cooks and gardeners, while staff inside the top-secret sector were 'almost entirely' American.

But was Boyce telling the truth about his espionage or was his claim to have been motivated by American duplicity towards Australia merely an attempt to save his own skin? According to Robert Lindsey, who reported on the trial, Boyce 'told his story ... crisply, without hesitation, and several times offered to take a lie-detector test'. Lindsey considered his testimony 'impressive'. But the court found him guilty and the judge, when passing sentence, rejected Boyce's 'attempt to rationalize his reasons ... to the effect that he was appalled at his government and what it was doing to Australia'. Doubting that there was 'any veracity within him', he sentenced Boyce to 40 years for espionage.

Des Ball notes that two years after the Boyce trial it was accepted by the CIA that geosynchronous SIGINT satellites controlled from Pine Gap could not be aimed at any particular targets without the knowledge of Australian technicians in the operations building and that Canberra had the authority to veto any target. Around this time Australians were allowed for the first time inside the station's most highly classified area, the signals analysis room. Four years later the senior Australian officer at Pine Gap was made chairman of the committee that met to decide on daily targets.

So how much harm, if any, did Boyce and Lee do by selling TRW's secrets to the Russians? In *Inside the Wilderness of Mirrors*, Paul Dibb states that information provided by the pair 'undoubtedly gave the Soviets significant operational insights'. He goes on:

> In my view, the Soviets understood—particularly, but not only, after the supply of operational parameters of Pine Gap by Christopher Boyce—just how much more advanced US intelligence collection was, not only with intercepted telemetry from ballistic missile testing and detecting their infrared signatures and locations but also of incredibly detailed access to Soviet tactical military operations (such as Nurrungar detecting Soviet fighter aircraft using their afterburners).

In *The Wizards of Langley*, Jeffrey Richelson quotes James V. Hirsch, who would later become the CIA's deputy director for science and technology, boasting that the agency 'knew more about Soviet telemetry systems than the Soviets did'. Hirsch may have been right, but Sputnik had showed that the Soviets were not always as backward as Washington (or Langley) liked to believe.

In July 1978, ten months after Boyce was sentenced in the federal district court, US satellites began picking up heavily encrypted telemetry signals from Soviet missiles. It was not the first time the Soviets had attempted to frustrate US intelligence analysts listening in on their telemetry transmissions. Four years earlier they had encrypted data from ICBM launches, but when the US protested that this contravened the terms of the first Strategic Arms Limitation Treaty (SALT), the encryption stopped. (Encrypted telemetry was against the spirit and letter of the SALT agreements, which required that each side should be able to verify what the other side was doing. Encryption by the Soviets theoretically made it impossible for the US to verify that Soviet missile activity complied with the treaty.)

When Soviet encryption resumed in 1978, it was (according to Des Ball) at 'much higher levels' than previously and was no longer confined to ICBMs. In his book *Endgame: The inside story of SALT II*, Strobe Talbott notes that while telemetry signals from the missile's booster phase had been encrypted before, this was the first time the Soviets had coded data about the performance of the warhead or re-entry vehicle. Such information was used by US analysts to determine the weight of the warhead and therefore the size of the missile's payload. This, Talbott writes, 'was information to which the US would feel entitled'.

According to an 'informed source' who spoke to the *Washington Post*, the encryption of signals such as missile telemetry was 'routine security practice'. A report headlined 'Spy satellites: "Secret," but much is known', published in January 1984, suggested that Soviet encryption of missile telemetry had nothing to do with Boyce's disclosures but was motivated instead by the demands of its new multiple warhead (MIRV) testing program. According to the *Post*:

> When Soviet ballistic missile testing involved only single-warhead missiles the United States was able to obtain the information it required via radar—hence, Soviet encryption would not have denied the United States any information of value. With multiple warheads, information about numbers of warheads, their design and accuracy could be obtained only from telemetry interception.

The Soviets tested another missile in December 1978. Again, American analysts quickly identified how much and what sort of data from the missile was being encrypted. By now both sides were locked in negotiations to finalise the terms of the next strategic arms limitation treaty, SALT II. The US secretary of state, Cyrus Vance, was due to meet his counterpart, Soviet foreign minister Andrei Gromyko, the day after the Soviet missile launch, where he would have been entitled, by the terms of the treaty, to raise the question of encryption. However, Vance faced a dilemma. By protesting to Gromyko about the Soviet use of encryption in the December test, Vance would have revealed how quickly US analysts had been able to detect and isolate the encrypted telemetry channels. This was information the Americans preferred to keep to themselves. In the event, Vance protested about the encrypted signals from the July test but said nothing about the signals from the December test. Vance later wrote that on 23 December 1978 he had been 'instructed to inform Gromyko that telemetry encryption as practiced in certain recent misile tests would violate the ban on deliberate concealment. Still, because it might reveal too much about our own intelligence capabilities, I could not explain what it was that was objectionable . . . Gromyko refused to respond to my statement.'

It is possible that the Soviets had faced a similar dilemma in 1975, when Boyce began passing information about the Rhyolite satellite program. From Boyce the Soviets learnt that the satellites were intercepting their telemetry signals and downloading the data to Pine Gap. Any sudden move to encrypt these signals risked tipping off the Americans that the Soviets had a source familiar with the CIA's geosynchronous satellite program. By July 1978 Boyce's treason had been discovered and Boyce himself was serving a long jail sentence, so the Soviets could safely encrypt their telemetry signals without fear of compromising an active source.

Was this what really happened?

In February 1980 the director of the CIA, Admiral Stansfield Turner, was asked by the House Foreign Affairs Committee, 'Is there any relationship between secrets sold to the USSR by a former TRW employee, Christopher Boyce, and the coding of Soviet missile telemetry?'

Turner answered, 'Probably not.'

Des Ball prefers the explanation that a change in the technology used by the Soviets to transmit signals made it easier to encrypt telemetry data.

What is certain is that Boyce and Lee betrayed the Rhyolite program to the Soviets, and that the increasing use of encryption after their betrayal undermined—and potentially neutralised—the role of Pine Gap and its geosynchronous satellites in the arms control process.

The ongoing problem for the Americans was that they could not confront Moscow with what they knew about Soviet telemetry without revealing how they knew it, and therefore making it easier for the Soviets to circumvent US detection in future.

During the next few years the Soviets were caught encrypting missile data at ever higher levels, from around 70 per cent to 98 per cent and even 100 per cent in tests of new ICBMs in 1982–83. The use of encoded telemetry was expanded from land-based to submarine-launched ballistic missiles and was used to conceal not only data relating to the booster and warhead but information about guidance, altitude and propulsion. The US Defence Intelligence Agency estimated that two Soviet ballistic missile launches in February and May 1983 were fully encrypted. In addition, both tests were carried out at night to hide the launch equipment from the cameras of US reconnaissance satellites.

The fallout from Boyce's spying activities inside TRW's black vault did not end with encrypted telemetry. The possibility that their coded signals could be intercepted and deciphered at a later date prompted the Soviets to begin jamming US geosynchronous satellites to block them from harvesting even encrypted data. Des Ball quotes testimony from the US assistant secretary of defense for international security affairs, Richard Perle, when he was asked by the House Foreign Affairs Committee about Soviet jamming:

> Question: I would like to know to what extent is the Soviet Union jamming the US satellites that are used for monitoring Soviet missile tests?

Mr Perle: There has been activity of the kind you suggest. It is very worrisome... There is a pattern, and a growing pattern, unhappily, of the Soviets resorting to a variety of devices that have the effect of denying us information that is critical to judging their performance under existing obligations. And interfering with satellites in the manner that you suggest is one of the devices that they have been resorting to.

The precise nature and effectiveness of the jamming, which took place only during missile flight tests, strongly indicated that the Soviets were making excellent use of technical data about the capabilities of the Rhyolite and Argus satellites controlled from Pine Gap—data they had been given by Christopher Boyce nearly a decade earlier.

So far the Soviets had been satisfied with encrypting their own signals and jamming US satellites trying to intercept them. Now they went a stage further, by reducing the power of transmissions from missile tests to such a degree they they could only be picked up by Soviet planes. As a result, the interception and downloading of useful missile telemetry data to Pine Gap virtually ceased.

Intercepted telemetry had been the primary means of monitoring the Soviet missile program since the 1960s. Much of this data had been routed through Pine Gap, enabling the Australian government to portray the CIA listening station as an indispensable piece in the jigsaw of international arms control, a keeper of the peace rather than a critical part of Washington's global military apparatus. By the mid-1980s even the US president, Ronald Reagan, was forced to admit to Congress that telemetry could no longer be relied on to verify compliance with arms control agreements.

Boyce's revelations did not make the CIA's geosynchronous satellite program suddenly obsolete, although they compromised its usefulness in some areas. Technical improvements by one side were (and are) inevitably matched if not surpassed by technical improvements by the other side. Larger antennas enhanced the satellites' ability to pick up low-power transmissions, while new generations of satellite launched in the mid-1980s were probably fitted with anti-jamming technology.

Boyce did not stay in jail for long. In January 1980 he escaped from Lompoc prison in California. One of America's most wanted men, he managed to elude US marshals and FBI agents for more than nineteen months. While sightings came from as far afield as Costa Rica and South Africa, Boyce was hiding much closer to home.

His fascination with peregrine falcons had drawn him to Port Angeles, around 130 kilometres from Seattle on the wild and mountainous Olympic Peninsula, one of the rare falcon's last remaining natural habitats. After a tip-off from an informer, police combed through applications for driver's licences in Washington state in search of someone who resembled Boyce. A man with the name Anthony Edward Lester looked remarkably like Boyce, and the handwriting on his application was a perfect match for Boyce's. Lester proved a hard man to find, and another tip-off revealed why. Boyce, alias Anthony Edward Lester, had robbed as many as sixteen banks in the north-west of Washington state during the previous year, getting away with amounts ranging from A$4000 to A$25,000. His face had been captured on a number of security cameras during bank hold-ups and he was lying low to avoid arrest.

Inquiries revealed that 'Anthony Lester' liked to spend time alone in the wilds of Washington, Montana and Idaho, often riding a mule. Nearly 30 marshals and FBI agents masquerading as loggers, hunters and other locals set out to find him, eventually cornering him in his car outside a drive-in burger bar in Port Angeles. According to a book called *Weird Washington: Your travel guide to Washington's local legends and best kept secrets*, two FBI agents 'pulled up next to Boyce's car and recognized him as . . . other agents surrounded the area. One of them yelled the now immortal warning, "Drop that hamburger!"'

Boyce was sentenced to 25 years for bank robbery and another three years for escaping from prison.

Chapter 18
Apocalypse now?

If nuclear war ever broke out between the superpowers, Pine Gap was certain to be a target. Both the US and Australian governments knew it. In a special report entitled *The Nuclear War Scare of 1983: How serious was it?*, published in 2013 by the Australian Strategic Policy Unit, Paul Dibb noted that 'Western spies for the Soviet Union' informed Moscow in the late 1970s about the intelligence and communications functions of the Australian bases. As vital parts of the US strategic apparatus, Pine Gap, Nurrungar and North West Cape were prime candidates for Soviet nuclear strikes.

On 8 December 1980 the director-general of the Office of National Assessments (ONA), Bob Furlonger, wrote to Sir Geoffrey Yeend, the secretary of the Department of the Prime Minister and Cabinet, about a highly classified study the ONA had undertaken on the instructions of the prime minister, Malcolm Fraser. 'Dear Geoff,' the letter began:

You may remember some time ago I told you that we were preparing a study on the effects on Australia of a nuclear war in the Northern Hemisphere. The subject is a complex one, and it has taken some time to finish the study. But it has now been completed and I am forwarding it today to the Prime Minister . . .

Apart from the Prime Minister's copy, the only other copies in existence are held by Bill Pritchett [Secretary of the Department of Defence], Arthur McMichael [director of the Joint Intelligence Organisation] and myself. I mentioned some time ago to Peter Henderson [secretary of the Department of Foreign Affairs] that we were doing the study, so that he would know of its existence, but he doesn't have a copy of it.

Yours sincerely,
Bob
(R.W. Furlonger)

The nuclear study was a collaboration between the ONA and the Joint Intelligence Organisation. Entitled 'A preliminary appraisal of the effects on Australia of a nuclear war', the paper was classified 'Top Secret AUSTEO'. The convoluted story of its handling testifies to the government's fear of the Australian public ever finding out about the study.

On 30 March 1981 the ONA delivered eleven copies of the paper to the deputy secretary of the Department of the Prime Minister and Cabinet, John Enfield. The copies were numbered from five to fifteen inclusive. Fraser wanted the paper to be discussed in the Foreign Affairs and Defence Committee of Cabinet, and copies were to be distributed to relevant ministers. Another four copies were later requested for distribution at the meeting, bringing the total to fifteen. A Cabinet office memorandum noted:

> Twelve copies were for the twelve Ministers on the FAD [Foreign Affairs and Defence] Committee. The other three copies were held for the Secretary and other Cabinet Officers present at the meeting for reference in case a detailed discussion of the paper ensued.
>
> The paper was circulated to members of the Committee in the Cabinet Room on 2 April and at the conclusion all copies distributed were recovered, sealed in a double envelope and placed in a limited access safe within the Cabinet Office strongroom. Normally such documents are returned to the originator as soon as possible after the meeting. However in this case as Ministers were to resume consideration of the paper at some later time it has been held in the Cabinet Office.
>
> On 13 April a copy of the document was given to Mr Enfield so that an analysis of the paper could be prepared for you before FAD Committee resumed its discussion. There has been no other access to the paper since the date of the meeting. The copy passed to Mr Enfield is still in his possession and has been sighted today. The remaining fourteen copies are held in the Cabinet Office as described above.

Despite the extreme security measures, it was not long before the press got wind of the government's nuclear study. On 26 July 1981 the *National Times* published an article by Marian Wilkinson under the headline 'Nuclear war: A 50-50 chance'. By now Bob Furlonger had been replaced as director-general of the ONA by Michael Cook. After reading the article Cook wrote a splenetic memorandum under the heading 'National Times article on nuclear war'. Like the original ONA paper, Cook's two-page memo was classified 'Top Secret AUSTEO'. The article, he wrote, 'makes five assertions about what is called "a study" [the very word, incidentally, used by his predecessor, Bob Furlonger, only without inverted commas] by ONA. Two of the assertions are about the substance of the study, two about its timing, and one about its circulation. All the assertions except the last are rubbish.'

Cook's purpose in writing to the prime minister was obvious. As head of the ONA, he wanted to refute any suggestion that the organisation had leaked. In her article Wilkinson did not name the study, although it was clear that she was referring to 'A preliminary appraisal of the effects on Australia of a nuclear war'. Her failure to mention the title was taken by Cook as proof that she had not actually read it but had 'at best simply heard, perhaps just surmised, that a study had been made and knew absolutely nothing more about it'.

The article, Cook wrote, made only 'two assertions of substance'. These were:

- 'There is now a 50 per cent chance that nuclear war will occur, according to a study by the Office of National Assessments (ONA), Australia's supreme intelligence co-ordinating body.'
- 'The ONA conclusion that the chances of a nuclear war have greatly increased . . .'

Both of these assertions exaggerated the ONA's findings, which came nowhere near estimating a 50 per cent likelihood of nuclear war. On the subject of timing, Wilkinson mistakenly wrote that the study had been done 'after the election of [President] Ronald Reagan'. Cook conceded that her assertion that the ONA paper 'had extremely limited circulation in Australian intelligence circles' was correct, but rejected her insinuation of a leak. 'The article is not the result of a leak,' he told Fraser. 'It is simply another piece of totally irresponsible journalism.'

Although inaccurate, Wilkinson's article needled a conservative government that, like its predecessors, was more concerned with keeping secrets than in acknowledging risks.

The ONA study was not the only one to explore the possibility of an Australian armageddon. The Joint Committee on Foreign Affairs and Defence and its Sub-Committee on Defence Matters were nearing the end of an eighteen-month examination of potential threats to Australia's security. Since calling for submissions through newspaper advertisements, the subcommittee had met 45 times, taking more than 1800 pages of public evidence as well as a large amount of 'in camera' evidence. The joint committee's report, entitled 'Threats to Australia's Security: Their nature and probability', was due to be tabled in November 1981. A month before it was tabled, an unsigned briefing paper canvassed issues that had been mentioned by Wilkinson in her *National Times* article and were likely to be raised in the joint committee's report.

Dated 20 October 1981 and headlined 'NUCLEAR FALL-OUT', the briefing paper posed two hypothetical questions—'Is Australia's planning for civil defence in the event of nuclear war adequate?' and 'What action does the Government propose to take?'—and suggested some 'possible' answers:

- Nuclear war between the superpowers is a possible but unlikely event.
- Australia is unlikely to be a target for nuclear weapons.
- Estimates of nuclear fallout on Australia can be derived from experience from nuclear testing between 1945 and 1962.
- In the case of total nuclear war confined to the northern hemisphere, radiation levels in Australia would rise by only about 50 per cent above the normal environmental level.
- The effects are difficult to estimate.
- There could be up to several hundred additional deaths per annum in Australia from cancer.
- Genetic effects are more uncertain but would be much less than carcinogenic effects.
- The massive economic, demographic and political changes in the northern hemisphere, in the aftermath of a strategic nuclear war, would pose much more serious problems for Australia than radioactive fallout.

Of the eight 'possible' answers, none made any attempt to address the questions. However, before these answers could be put to the test, the Joint Committee on Foreign Affairs and Defence tabled its long-anticipated report.

Like other parliamentary committees bent on prying into the secrets of Pine Gap, the Joint Committee on Foreign Affairs and Defence was frustrated by Australian government secrecy, noting in its report that most of what was known about Pine Gap came from 'official and unofficial sources in the United States'. Again, like other committees, it turned to Des Ball for information it could not obtain from the government. Ball made it clear that the function of Pine Gap went far beyond arms control and 'counting numbers of missiles in the Soviet Union or the numbers of radars or whatever', but was used for locating them and 'allowing more accurate targeting in the development of American nuclear war fighting capabilities'.

There was 'no doubt', Ball told the committee, about the 'value of intelligence' obtained from the joint facilities.

> In the case of the infra-red intelligence [i.e. early warning of missile launches] ... Australia is only one of two or three places where this intelligence is passed down to the ground, the other one being in Buckley, Colorado. In the case of signals intelligence there have been some of what I think are probably the biggest intelligence breakthroughs of the late 1970s; listening in on Soviet microwave communications ... was probably the biggest technical intelligence coup ... of the 1970s. And again, that had to be done from Australia for technical reasons.

Guided by Ball's account of its function and capabilities, the joint committee concluded that 'it would be prudent' for Australian defence planners to assume that Pine Gap was on the Soviet target list and 'might be attacked in the course of a nuclear conflict between the two superpowers'. Rather than expend one of its multi-warhead SS18 ICBMs on a 'soft' target such as Pine Gap, the committee expected the Soviets to use the older, less destructive and 'relatively inaccurate' SS11, set to detonate at an altitude of about 900 metres. Using these assumptions, the committee predicted that nuclear strikes on the three US bases would cause the following 'immediate damage':

a. North West Cape: complete destruction of the communications station and the nearby town of Exmouth;

b. Pine Gap: complete destruction of the facility; marginal damage (broken windows, small fires, etc.) to Alice Springs (which is approximately twenty kilometres away);
c. Nurrungar: complete destruction of the facility; damage to windows, tiled roofs and wooden buildings plus 'spot' fires in Woomera Village.

Drawing on a seven-page booklet published in 1964 by the Commonwealth Directorate of Civil Defence called 'Survival from nuclear attack: Protective measures against radiation from fallout', the committee asserted that an air burst (as opposed to a ground-level detonation) would minimise radioactive fallout. Consequently, casualties would be confined to the three facilities and 'nearby inhabited areas' but these casualties would be 'drastically reduced' if the opportunity were taken during the lead-up to nuclear war to evacuate 'non-essential people' from the facilities as well as from Exmouth and from Woomera Village. Evacuation of Alice Springs 'would not be necessary'. Relatively simple precautions in Alice Springs and Woomera Village, such as 'whitewashing and taping windows, installing shutters, cleaning up combustible material and constructing simple shelters' would significantly reduce non-lethal casualties that might be caused by heat or collapsed roofs.

The report was tabled in the House of Representatives on 18 November 1981 by the chairman of the subcommittee on defence matters, Bob Katter, a Queensland senator for the National Country Party and the father of the current federal member for Kennedy. He hoped it would provide a 'framework for public discussion'. His colleague Bill Morrison, Whitlam's former defence minister, told the parliament that the purpose of the report was to 'provide a basis for a sane and sensible debate'.

The press showed some interest in the joint committee's conclusion that North West Cape, Pine Gap and Nurrungar would be on a list of Soviet nuclear targets. However, the Fraser government was no more inclined to encourage well-informed public debate about the US bases than any of its Coalition predecessors. While Fraser was powerless to suppress a report that had been tabled in the parliament, he was determined to keep the ONA paper safely under lock and key. The Australian public remained oblivious to the study's glib acceptance of the 'possibility, given Soviet war-fighting doctrine—which places a high value on pre-emption—that the US facilities in Australia might be targeted relatively early in a strategic nuclear war'.

> As the nuclear conflict escalated and the prospects of its containment receded, we judge that nuclear attacks on some or all of these facilities would probably occur. The United States deploys about half its nuclear warheads in submarine-launched missiles. Therefore if the United States was using, or was judged likely to use, its submarine forces to strike at the opposition's cities, the USSR would rank North West Cape as an important nuclear target.

While the ONA paper found the likelihood of nuclear strikes against Pine Gap and Nurrungar 'may be somewhat lower', it noted that such attacks 'cannot be excluded'.

In fact, statements made by Soviet military commanders clearly indicated that not only could the possibility of attacks against Pine Gap and Nurrungar not be excluded, but according to Soviet military orthodoxy going back to the early 1960s they were virtually guaranteed. Doubters needed to look no further than 'Annex A' of the same ONA study, which under the heading

'Soviet doctrine on the nuclear nature of world war' quoted the following extract from Marshal Vasily Sokolovsky's 1968 book *Military Strategy*, described as 'one of the basic documents of Soviet military doctrine':

> The logic of war is such that if a war is unleashed by the aggressive circles of the United States, it will immediately be transferred to the territory of the United States of America.
>
> Those countries on whose territory are located military bases of the United States, NATO, and other military blocs, as well as those countries which create these military bases for aggressive purposes, would also be subject to shattering attacks in such a war. A nuclear war would spread instantly over the entire globe.

It was wishful thinking to imagine that Australia was not one of those countries. In the event of a nuclear war involving Australia, the ONA paper observed that

> attacks on the Joint Facilities using ground or low-altitude detonations (ground bursts) are less likely that higher altitude detonations (air bursts), which produce little local fallout; the most militarily effective form of attack on unhardened facilities such as those at Pine Gap, Nurrungar or North West Cape is generally an air burst attack, perhaps involving detonation of about a one megaton warhead [about 75 times the yield of the Hiroshima bomb] at an altitude of about 1,000–1,500 metres above the target. An explosion at this altitude would maximise the area of blast damage on the ground, and hence minimise the effect of targeting inaccuracy . . .

In the event of such attacks using single air bursts against Pine Gap and Nurrungar, and assuming no unusual targeting error, the neighbouring towns of Alice Springs (population 14,000) and Woomera (population 2,000) would not suffer major damage; there would be broken windows, the hazard of flying glass and other debris, and isolated fires. Casualties would be strongly dependent on the extent of civil defence measures. There could be some fatalities. Following any type of nuclear attack on North West Cape, the town of Exmouth (population 2,000) would probably be destroyed. Most of the radio-active debris from these air bursts would rise into the stratosphere, where it would encircle the globe. Local fallout would probably not be a major problem.

The ONA paper had the feel of a document researched and written in Canberra, to be read in Canberra. Nuclear war looked different when contemplated from Alice Springs.

In 1985 the Medical Association for the Prevention of War (NT) and Scientists Against Nuclear Arms (NT) published a booklet called 'What will happen to Alice if the bomb goes off?' The author was a local doctor, Peter Tait. Drawing heavily on the work of Des Ball, Tait's booklet was in many ways the antithesis of the ONA study, highlighting phenomena (such as the 'flash' from a nuclear explosion) that had been minimised or ignored altogether by the authors of the ONA paper but would have a devastating effect on anyone caught in the blast.

The flash would blind anyone up to a distance of 80 kilometres away who was looking in the direction of Pine Gap at the moment of the explosion or who, noticing the flash, instinctively glanced towards it. People in the south and western town area are shielded from the flash by the ranges. People on the East Side and outside Heavitree Gap, however, would be likely to see the flash.

Heat was another phenomenon largely neglected in both the joint committee's report and the ONA paper, despite the fact that in the Hiroshima and Nagasaki explosions thousands of victims were incinerated. While much of Alice Springs was far enough from Pine Gap to escape the intense heat of an air blast directly above the base, some suburbs were much closer. In his booklet, Dr Peter Tait clearly identified the vulnerable areas:

Effects of the heat released by the explosion extend to about 15 to 20 kilometres. Similar to effects from the blast, most damage would occur outside Heavitree Gap.

Everything flammable within 10 kilometres of Pine Gap would catch fire. This includes the White Gum Estate, parts of the airport road and Stuart Highway, a section of Larapinta Drive, the Rangers' Station at Simpsons Gap and all scrub and animal life in that 10 kilometre radius.

Any people in the open along the South Road, at the airport and even at Simpsons Gap would receive the equivalent to mild up to severe sunburn, depending on exactly where they were. Any people in the open west of the South Road would receive third degree burns.

Like the joint committee and the ONA, Peter Tait worked on the premise that a Soviet nuclear attack on Pine Gap would come in the form of a one-megaton 'air burst' roughly 900 metres above the centre of the base. Detonating a warhead at this altitude would cause maximum blast damage. But what if Soviet military planners chose to attack the bases with 'ground bursts', sacrificing blast damage in favour of soaking nearby civilian populations with lethal levels of radiation? According to the ONA study, strikes on either Pine Gap or Nurrungar would cause massive fallout from dirt picked up by the explosion. If the wind direction were within about 45 degrees of Alice Springs and Woomera, 'most people not evacuated within one hour would receive fatal radiation doses'. Protective shelters, or procedures for rapid evacuation from the danger zone, would therefore be 'vital'.

As a local, Peter Tait knew that such an outcome was relatively unlikely, since the wind in central Australia usually blows from the southeast. However, 'if on the day [of a nuclear strike] it was blowing from anywhere in the southwest quarter, Alice Springs would be enveloped in the plume of radiation at greater than the lethal dose . . . This would happen within hours of the explosion. Everyone would die of radiation poisoning within 24 hours . . . blast and radiation effects would stress the medical services in central Australia beyond coping. Very many people would die, untreated. Large tracts of central Australia would become uninhabitable.'

The danger from ground-burst attacks would not be confined to nearby population centres. The ONA study predicted that dangerous amounts of local fallout would be carried hundreds of kilometres by the wind. Fallout from a low-altitude detonation at Nurrungar 'might threaten South Australian coastal towns . . . northerly high-altitude winds, which occur about five per cent of the time, could expose the inhabitants of the Port Augusta, Whyalla, and/or Port Pirie areas to high radiation levels which, if evacuation did not take place, would amount to a fatal dose within a few days ... Depending on wind speed, only a few hours would probably be available for evacuation before the arrival of fallout unless there was prior warning of a likely Soviet attack.'

Adelaide, 400 kilometres south of Nurrungar, was considered relatively safe, although in the 'unlikely' event of a northerly wind carrying fallout over the city, high radiation levels 'could justify temporary evacuation'. Failure to evacuate the city, according to the ONA study, could cause radiation deaths and 'would, in the long term, perhaps cause about one thousand additional cancer deaths and a detectable increase in genetic defects'. The evacuation of hundreds of thousands of people from Adelaide would, it noted, 'pose considerable problems of organisation and communication'.

The ONA paper presented a scientifically plausible vision of what a nuclear strike on Australia might look like, but concluded that such an attack was 'unlikely'. The risk of a nuclear war in the northern hemisphere was also thought to be low. '[N]othing in this study should be read as implying that nuclear war is probable,' the authors wrote. '[W]e consider that the formidable constraints on nuclear war between the superpowers will probably continue to be effective.' The ONA was confident enough in its assessments to state that they were 'valid for at least five years and, in broad terms, probably for ten years'.

Not everyone shared that confidence. Asked for his opinion of the ONA study, Paul Dibb recalls advising the secretary of the defence department that 'if I were a member of the Soviet General Staff I would suspect that the US bases in Australia ... did support an offensive nuclear war-fighting system that assisted the Americans more accurately to target the location of Soviet ICBMs and their destruction in a nuclear war.

Soviet paranoia and what they understood from their spies in the United States and the Soviet Embassy in Canberra would mean that Australia was almost certainly targeted and that the joint facilities would rank high in Soviet nuclear war-fighting priorities. And that might also mean targeting major Australian cities such as Sydney and Melbourne.'

Dibb, a former deputy director of the Joint Intelligence Organisation, had not contributed to the ONA–JIO study, but his wide knowledge of Soviet affairs, as well as his personal contact with Soviet diplomats, made him a shrewd judge of Soviet strategy.

The world was closer to nuclear war in the early 1980s than at any time since the end of the Second World War—closer, arguably, than it had been during the tensest moments of the 1962 Cuban missile crisis. Under its hawkish new president, Ronald Reagan, the US had adopted a more belligerent anti-Soviet posture. The Pentagon was developing or deploying new weapons systems, including the 'Star Wars' missile defence shield, that appeared to represent an existential threat to the Soviet Union. In speeches Reagan referred to the USSR as the 'evil empire' and vowed to consign it to the 'ash heap of history'. Oleg Gordievsky, a KGB officer turned British double agent, revealed that the Moscow leadership was terrified of a possible pre-emptive nuclear attack by the West. During NATO war games in 1983 the Soviets went on high nuclear alert, fearing an imminent strike.

Washington could not conceive that its aggressive war-gaming (which included loading planes with dummy nuclear weapons) was being read in Moscow as preparation for a real attack, but Gordievsky confirmed that this was exactly how the Kremlin was interpreting them. An alarmed Reagan began to look for other ways to bring the Cold War to an end, and the ONA's breezy assertion that there would be no nuclear war between the superpowers turned out to be right.

The ONA paper was shown to select ministers and discussed at a meeting of the Cabinet's Foreign Affairs and Defence Committee on 2 April 1981. Fraser discussed the paper afterwards with Mike Codd, undersecretary at the Department of the Prime Minister and Cabinet, and his colleague John Enfield. According to a now declassified memorandum written by Codd, there was to be another FAD meeting at which ministers would be 'given another opportunity to consider the matter—with a reading period for a quarter of an hour ahead of the meeting'.

In his own assessment of a prime ministerial 'brief' he considered 'surprisingly long-winded', Dibb declared the ONA's analysis 'far too relaxed about the Soviet nuclear threat to Pine Gap, Nurrungar and North West Cape'. He considered that major Australian cities were 'at greater risk' than had been portrayed and that '[t]hroughout the ONA assessment there was . . . a tendency to underrate the catastrophic effects of nuclear war'.

A series of scribbled questions—'What is our civil defence plan? What is being done in the US? the UK?'—at the bottom of an archived copy of Codd's memorandum suggests that he shared Dibb's concerns. The bureaucracy, however, was preoccupied with keeping the ONA study secret. 'Because of the sensitivity of this subject,' the ONA chief, Bob Furlonger, told Enfield, 'I would be grateful if all copies of the paper could be returned to this office for destruction after the relevant meeting [of the Foreign Affairs and Defence Committee].'

It took nearly two years for Furlonger's instructions to be carried out in full. A Cabinet Office document entitled

'Certificate of destruction of classified matter' records that on 10 March 1983, fourteen copies (numbered six to nineteen) were destroyed in front of a witness and that copy number two had been 'sent to ONA for their records'. The ONA paper was itself classified 'Top Secret AUSTEO' and so was the certificate confirming its destruction. Another copy, number five, was described as being 'retained on a Cabinet file for record purposes'. This is the copy that found its way into the National Archives and was released in 2013 under the *Freedom of Information Act*. More than three decades after it was written, parts of the paper remain heavily redacted.

Chapter 19
Reds, ratbags and radicals

On 19 October 1977, the eleven-year-old Pine Gap agreement between the US and Australian governments was extended by an Exchange of Notes. Once the new agreement had been in force for nine years (i.e. after 19 October 1986) it could be terminated by either party with one year's notice. Officially, the base was now secure, at least for the next decade. But the voices questioning its usefulness, its safety and the morality of its existence were growing louder.

For many years, the best informed and most troublesome of the base's critics had been Des Ball and his colleagues at the Australian National University. According to Paul Dibb, Ball's increasingly prolific writings about Pine Gap were viewed by the defence department, and especially its bulldog secretary, Sir Arthur Tange, as 'aiding the enemy'. When Dibb left the department to work at the ANU on his book *The Soviet Union: The incomplete superpower*, Tange's successor, Bill Pritchett, told him, 'You'll be working along the corridor from that Des Ball and we'll be watching you.'

Dibb described the department's obsession with Des Ball as 'paranoia', but its conquences were real enough. The National Archives has two ASIO files on Ball dating back to the time when he was a student at ANU. Some of the material consists of newspaper cuttings but the files also contain detailed reports, many stamped 'Secret' and partially redacted, that show how Ball had been kept under ASIO surveillance at least since 1966, when he was charged with 'offensive behaviour' during an anti-Vietnam war demonstration in Canberra.

As a final-year political science student, Ball had aroused further suspicion through his collaboration with Robert Cooksey on an article about Pine Gap published in the *Australian Quarterly* in December 1968. A five-page letter written in April 1969 by the then director-general of ASIO, Sir Charles Spry, in response to a 'verbal request' by Tange's predecessor, Sir Henry Bland, indicates how seriously the government and ASIO took the activities of Ball, Cooksey and other Australian academics interested in 'defence installations and defence policy', many of whom were alleged to 'have publicly known radical sympathies and ... contact with the Communist Party of Australia'.

In 1969 Cooksey was a lecturer in political science in the ANU's School of General Studies. Ball would go on to become head of the ANU's Strategic and Defence Studies Centre, which in December 1968 had been the subject of an article in the communist *Australian Left Review*. The article, by Monash University lecturer John Playford, was described by Spry as having been written 'with assistance from sources which are not normally available to academics'. In his letter to Bland, Spry speculated that Playford had been 'acting as a disinformation agent on behalf of the Russian Intelligence Service'.

Noting the number of Australian academics with 'radical'

left-wing political sympathies writing about Pine Gap, the ASIO chief advised Bland that their activities were intended 'to embarrass the Commonwealth Government in its field of defence research and analysis and also to isolate and embarrass any academics who have contact with your Department, and . . . to ensure that the numbers of such people are limited'.

The linking of 'communists' with Pine Gap rang alarm bells in the security services of both Australia and the United States. Ball never doubted that Bland's request for information on him was prompted by the CIA's then chief of station in Australia, Raymond Villemarette. In *A National Asset: 50 years of the Strategic and Defence Studies Centre*, Ball writes that the CIA 'was concerned that my research might reveal both its role and the existence of its geostationary SIGINT satellite program'. (In the same essay Ball admits that he did not know the CIA was in charge of Pine Gap until Brian Toohey revealed it in November 1975 in the *Australian Financial Review*.)

Spry's suspicions of Soviet infiltration into the ANU's Strategic and Defence Studies Centre were not as outlandish as they seemed. In 1999 *News Weekly*, published by the National Civic Council, alleged that 'for over two decades, the KGB has regarded SDSC as a key target area in which they can recruit agents of influence' and gain access to agents 'in Defence and the intelligence community'. In his essay in *A National Asset*, Des Ball recalls that Lev Sergeyevich Koshlyakov, the 'energetic KGB Resident in Canberra from 1977 to 1984', was said to have been 'well known to key senior Centre staff', although Ball did not believe he ever visited the centre. Another Soviet diplomat, Igor Saprykin, whom ASIO suspected of being a KGB officer, did visit the centre. But the KGB was not alone in trying to cultivate the staff of the Strategic and Defence Studies Centre.

According to *News Weekly*, two of the centre's 'directors' were ASIO sources. It is likely that at least one of the pair was feeding Spry information about Cooksey and Ball.

In May 1969, a month after Spry's letter to Bland, the ASIO chief asked his Victorian regional director to conduct a 'birth check' on Ball. (This was later supplied.) The following month Ball took part in a noisy protest against the visit of the South African minister for economic affairs, Jan Haak. An ASIO report lists Ball and Cooksey among the demonstrators ('a few ... with their faces painted black') standing outside the Hotel Canberra and 'shout[ing] slogans such as "Go home Haak—take Gorton with you"'.

On 2 July 1969 the *Sydney Morning Herald* and the Melbourne *Age* published the first of a series of articles about Pine Gap by Cooksey and Ball. Under the headline 'Pine Gap: What is its purpose?' the two authors described what they considered to be the 'two ongoing functions of Pine Gap', namely the reception of information from reconaissance satellites (especially early warning satellites) and the control of space-based anti-ballistic missiles.

In the second article, published the following day, Cooksey and Ball argued that Pine Gap would be a 'priority target' in a global nuclear war and that 'major cities, especially Sydney and Melbourne' would also be hit by Soviet missiles.

> It is more than possible that the Soviet Union might become aware ... of a successful development program of satellite-ABMs from Pine Gap. Then it might issue an ultimatum to the Australian Government to dismantle the installation, or suffer a nuclear attack on an Australian city. Or perhaps it would launch a preventive [sic] strike against Pine Gap.

In the final article, 'Real reason for veil of secrecy', Cooksey and Ball questioned the value Australia was getting from the US alliance and argued that the ultimate beneficiary of Australia's hosting of Pine Gap was the United States. 'Under Australian secrecy,' they wrote, 'American Administrations can undertake projects which are difficult to handle at home.' The article concluded:

> In any event, Pine Gap is operational and likely to remain so. As have other peoples, Australians will no doubt learn to live with the prospects of nuclear war.
>
> But it is no comfort that the Government took such a critical decision as agreeing to the construction of the Pine Gap installation, privately and probably ill-advisedly, without consulting Parliament or allowing public debate.
>
> Nor does it help that Pine Gap serves only the global strategic interests of another country—even if that country is the USA.

Although the three articles ran to a total of several thousand words, they contained relatively little hard technical data, and probably nothing that was not already available from published sources in the United States. But Cooksey was preparing to raise the stakes.

A declassified document in Ball's ASIO file reveals that on the evening of 10 September 1969 a source inside the ANU informed ASIO that Cooksey had announced his attention to 'try and visit the Pine Gap installation later this week'. This is the visit (described in chapter 8) that Cooksey would later write about in the *Age*. The then opposition leader, Gough Whitlam, had visited Pine Gap a week earlier but had been prevented from entering the technical areas. Cooksey had submitted a written request to the defence department to be allowed to visit but permission was refused.

According to the ANU informant, Cooksey planned to leave Canberra on 10 September before travelling to Alice Springs via Melbourne and Adelaide. His 'exact mode of travel' was not known, but Cooksey had made it known 'in University circles' that 'funds had been made available for his trip'. The informant did not know the source of the funds but considered it 'most unlikely' that the money came from the ANU.

From the moment he arrived in Alice Springs, Cooksey was under surveillance. The departments of defence and supply had been warned of his arrival and a defence official, Mr F.S.B. Appleton, DSO, was on hand to confirm that the ANU lecturer landed at Alice Springs airport 'at about midday' and checked into a local hotel. It was also learnt that 'he was using an Avis rent-a-car and the suggestion was made in press circles that funds had been made available to him by the "Australian" [newspaper]'.

The ANU informant had picked up some talk around the university that Des Ball was planning to accompany Cooksey (in fact Cooksey made the trip alone). ASIO's assistant regional director for the ACT believed the plan was 'to attempt to gain entry into the Joint Defence Research Station at Pine Gap'. At 1000 hours on 15 September:

Mr [assistant regional director, ACT, name deleted] . . . stated that the most recent information concerning COOKSEY was that he had spent the weekend prowling around Pine Gap. He made several unsuccessful attempts to enter the area, including an attempt to crawl under the large wire mesh fence. The air charter services in Alice Springs had also refused to fly him over or near the area. The question of financing . . . [the] trip had been raised in Canberra, and it had been suggested that 'The Age' may be providing the money.

In what was described as an 'encounter' with a Commonwealth Police officer at the eastern boundary fence of the joint facility, Cooksey gave his name and also 'mentioned the name of Desmond John Ball when speaking of the articles that have been published in the Melbourne Age . . . This was the only time that Ball was mentioned.'

The documents in Ball's ASIO file suggest that during his visit to Alice Springs, Cooksey was followed everywhere he went. While in town he asked for the Lands Office to be specially opened so that he could buy maps of the Pine Gap area. According to a report, the Lands officer 'met Cooksey at the Lands office and showed him a copy of the normal Alice Springs District Map scale 1:250,000 produced by National Mapping which is for public sale . . . at this time a copy was not available for sale, and the Lands officer was not agreeable to recalling a staff member to duty at that hour to produce a black and white print of same. This map shows the access road from the South Road to the Pine Gap entrance to the Prohibited Area. It does not indicate the boundaries of the Prohibited Area.'

Thwarted in his efforts to get inside the installation, Cooksey left Alice Springs by TAA Flight 555 at 12.45 pm on 14 September, apparently because of 'lectures he had to deliver at the ANU on the afternoon of 15 September'. After the encounter with the policeman Cooksey reportedly said that he had 'no intention of making a martyr of himself by illegally entering the

JDSRF Prohibited Area after permission had been refused by the Minister for Defence'.

As an individual trying to get a close-up view of Pine Gap, Robert Cooksey was a nuisance but nothing more. Of much deeper concern for ASIO was the possibility that he was acting on behalf of (and perhaps being funded by) a radical political group. While Cooksey and Ball had both taken part in organised political protests against the Vietnam War and against South Africa's apartheid regime, Cooksey was not believed to be a member of the Communist Party. According to a declassified ASIO document he regarded himself as being 'left wing ALP' and had stated that he was 'not a communist and would not entertain the thought of joining the Communist Party' because of the 'irrelevancy of Marxism to Australian conditions'.

During his brush with the police officer outside the boundary fence, Cooksey 'did not mention his association with any organisation other than the ANU [and] "Age" newspaper ... the opportunity did not occur ... to glean any further information regarding Ball's membership of any associations'.

In his letter to Sir Henry Bland, the ASIO chief described Ball as 'an associate of students who have radical views', while Cooksey had been 'reported as having contact with Communist Party of Australia personalities' and was 'considered to be indiscreet'. Try as it might, ASIO could find no evidence that either Cooksey or Ball was a member of the Communist Party. Ball, however, would continue to be viewed with suspicion by senior officials in both defence and intelligence, who regarded his efforts to prise open the secrets of Pine Gap as tantamount to treason.

It was not only Arthur Tange and the heads of ASIO who distrusted Ball. Between 1975 and 1980 Milton Corley Wonus was the CIA station chief in Canberra. Wonus was an expert in signals intelligence, having worked as a US Air Force SIGINT analyst in Japan during the 1950s. Unlike previous CIA station chiefs, Wonus had a background in the agency's Directorate of Science and Technology, the division responsible for the design and operations of the Rhyolite spy satellite program and the development of Pine Gap.

The American was not an admirer of ASIO. According to Paul Dibb, Wonus felt that ASIO 'viewed the Russians from the standpoint of the 1950s', that it had a 'blinkered view of Soviet operations in Australia' and was still behaving like 'the policeman on the beat'. Whereas Wonus believed the Russians were playing a 'sophisticated and subtle game, especially in . . . seeking political and other influence in high places', ASIO was stuck in the past and spent its time 'tailing the Russians on their more obvious contacts with Australian communists'.

Dibb wrote that Wonus had 'two pet hates'. One was the US spies Boyce and Lee, whom he 'wanted to see . . . executed'. The other was Des Ball, who, in the American's view, had 'revealed dangerous information about US intelligence installations in Australia, especially Pine Gap'. Fortunately for Ball, the CIA station chief expressed no desire for him to be terminated.

Chapter 20
No admission

On 22 March 1967 Pine Gap's first chief of facility, Richard Stallings, assured Alice Springs residents that there would be 'no serving military officers or men at the site'. For the next two decades, Australian governments of both political stripes conspired in the fiction that Pine Gap was not a military base.

By the late 1980s, as Des Ball pointed out in the preface to *Pine Gap*, it was clear that the signals intelligence harvested in central Australia contributed more to the Pentagon's nuclear war-fighting apparatus than it did to the international cause of global arms control, and that, in the event of a nuclear conflict between the superpowers, Pine Gap—and the unprotected civilian population of Alice Springs—would be a prime target.

Whether there were 'serving military officers or men' at the site or not, Pine Gap was a vital part of the Pentagon's global operations, and a linchpin of US national security. A string of Australian prime ministers had guarded its secrets—in Whitlam's case, even as he raged against the Nixon administration and murmured threats against the bases—but after Malcolm Fraser the commitment to secrecy fractured.

In March 1981 the Labor leader, Bill Hayden, and his deputy, Lionel Bowen, requested and were given detailed confidential briefings on the joint facilities and were allowed to visit the bases at Pine Gap, Nurrungar and North West Cape. The aim of the visits, Hayden told a National Press Club lunch in Canberra, was to determine whether 'the existence and function of these facilities was consistent with the principles of unimpaired Australian authority and sovereignty'.

In his speech Hayden openly challenged the 'obsessive' secrecy that surrounded the bases, arguing that it made Pine Gap and Nurrungar appear 'unnecessarily sinister'. The goal of the Labor Party in government, he said, would be to 'modify those restrictions'.

While he was forbidden to talk about what went on inside the bases, Hayden claimed that Labor leaders had been given 'full access' during their visits and that intelligence officials had supplied them with 'detailed information on the purpose and operations' of the joint facilities. As a result of their briefings, he and Bowen were 'satisfied' that Pine Gap was 'not part of any weapons control system' and 'cannot initiate or control any military operation'. Rather, he said, they made an important contribution to 'strategic nuclear stability and deterrence'.

As to the question of Australian 'authority' over the base, Hayden told his audience that Australia had 'complete access to all the data', that the data was 'monitored by Australians' and that any attempt (by the Americans, presumably) to contravene the agreed terms of operation 'would immediately be noticed by Australians'.

The Hawke Labor government, elected in a landslide on 5 March 1983, was determined not to repeat the mistakes of the Whitlam government a decade earlier. Unlike Whitlam, whose relations with Nixon were marked by a mutual antagonism verging often on contempt, Hawke got along well with Republican presidents Ronald Reagan and George H.W. Bush. Not everyone in the Labor caucus shared his enthusiasm, but Hawke's electoral appeal gave him the authority he needed to keep the party united (albeit uneasily at times) behind the US–Australian alliance. By the mid-1980s, however, absolute secrecy no longer seemed the best way to protect the alliance's most controversial and recognisable asset: Pine Gap.

On 6 June 1984 Hawke made a speech on arms control and disarmament in the House of Representatives. In his statement he drew attention to Australian efforts on behalf of the nuclear non-proliferation treaty (NPT) and a treaty outlawing the testing of nuclear weapons. The Australian government, he said, regarded the NPT as 'the most important multilateral non-proliferation and arms control agreement in existence' and would do everything in its power to help enforce the treaty. More than a decade earlier, opponents of the NPT had urged its rejection on the grounds that Australia 'would need a nuclear weapons capacity to repel the survivors of an atomic war in the Northern Hemisphere'. The refuseniks had been defeated, and Australia ratified the NPT in January 1973. Hawke reiterated that his government 'categorically rejects any nuclear weapons option for Australia'.

The same year, Hawke's foreign minister, Bill Hayden, stated publicly that the primary function of Pine Gap was 'monitoring as part of verification of compliance with the provisions of arms control agreements' and that it was 'highly unlikely that some major arms control agreements between the superpowers would have been concluded if there had been no Pine Gap'.

But the Pine Gap station had grown, in size and capability, and its role now went far beyond arms control. Hayden's remarks about the value of Pine Gap to arms control verification were not false, but they gave a misleading impression of the spy station's broader surveillance functions.

Since the construction of the first two antennas/radomes in 1968, the number had mushroomed. Two more were built in 1969 (one of which was replaced in 1973) and another four were installed between 1971 and 1985. A ninth radome would appear in 1989, and two more in 1990–91. Between 1970 and mid-1990, staff at the base increased by 50 per cent, from 440 to more than 650.

Among other things, Pine Gap hosted key satellite ground terminals for the US Defense Satellite Communications System (DSCS), described by Des Ball in his 1992 paper 'Defence aspects of Australia's space activities' as 'perhaps the largest single US military satellite program'. While the GPS navigational/positioning satellite system boasted more satellites operating at any given time, the DSCS program generally commanded the largest annual budget, used the largest number of major satellite ground terminals around the world and, according to Ball, supported 'an extremely wide range of US global military and intelligence missions'.

The authoritative *Jane's Military Communications* describes the DSCS as having been 'designed and configured for presidential communications'. It supported the Worldwide Military Command and Control System by providing communications between the national command authorities, unified and specified commands, and general combat forces; it provided communications from 'peripheral early warning sites' and intelligence sites in support of 'contingency and limited war operations'; it provided wide-band channels needed to handle 'high-quality secure voice [and] high-speed data between automated command and control centres'; and it supported 'Navy ship-to-shore communications'.

According to Ball, the DSCS consisted of eight geosynchronous satellites operating from four positions (over the Atlantic Ocean, East Pacific, West Pacific and Indian Ocean) and linked to six Australian satellite ground terminals: one at North West Cape; two at Pine Gap; one at Nurrungar; one at Watsonia Barracks in Melbourne; and one at the Weapons Research Establishment at Salisbury, South Australia.

The principal mission of the Pine Gap station—still referred to in US intelligence cables by its unclassified National Security Agency name, Rainfall—was the same as it had been since the days of Rhyolite: the collection of signals intelligence from US spy satellites in geosynchronous orbit over the equator. Controlled from Pine Gap, the satellites intercepted a huge range of electronic data which was then downlinked for processing and analysis.

The first generation Rhyolite satellite, 'Bird 1', had been superseded by more powerful satellites with larger and more sensitive antennae. On 24 January 1985 the space shuttle *Discovery* took off from the Kennedy Space Center in Florida with a new satellite on board, initially designated Magnum-1 but later renamed Orion-1. Like the Rhyolites, Orion-1 had been designed and built by TRW Inc. of Redondo Beach, California, and would be controlled from Pine Gap.

Rhyolite-1 reportedly weighed 700 kilograms with a full fuel load. Orion-1 was much larger and, at around 2500 kilograms, around three and a half times heavier. At the time of its launch, Orion-1 was described as the largest spacecraft ever placed in geosynchronous orbit.

The Pentagon and the CIA went to unprecedented lengths to keep details of the new satellite secret, but their efforts backfired spectacularly. A month before the launch, the US defense secretary, Brigadier General Richard Abel, chief of air force public affairs, had threatened to investigate anyone who even 'speculated' about the content or capability of the shuttle's secret military operations. At an extraordinary press briefing Abel told reporters, 'We are working to deny our adversaries any information that might reveal the identity or missions of our DOD [Department of Defense] payloads.' At the same time the US defense secretary, Caspar Weinberger, warned the three US television networks, the Associated Press and a trade magazine, *Aviation Week & Space Technology*, that disclosure of the shuttle's payload could 'seriously endanger national security'.

Under the headline 'Spy satellites: "Secret," but much is known', *Washington Post* reporters Jeffrey Richelson and William Arkin revealed on 19 December 1984 that the space shuttle *Discovery* was to deploy a 'new military intelligence satellite that is to collect electronic signals and retransmit them to a U.S. receiving station on Earth'. They described the $300 million satellite as 'the most important and largest' of 'four or five' already in space. Weinberger was livid, accusing the *Post* of 'helping' the Soviet Union and harming 'national security'.

Affronted by Weinberger's attack, the *Post* reacted by playing down its own scoop, insisting that 'the Soviets were well aware of our ability to collect signals from space' and pointing out that the facts contained in the article were 'easily deducible from information available in congressional testimony, *Aviation*

Week, *The New York Times*, official Australian announcements and other unclassified sources'.

The suggestion that the Soviets could gain a strategic advantage in space by reading the *Washington Post* was, the paper declared, a 'pipe dream'. The United States and Soviet Union, it said, 'both maintain vast networks that collect information about the other's eavesdropping activities. Each superpower's space monitoring networks include detection and tracking radars, telescopes that can see out to geosynchronous altitudes, ships equipped with electronic eavesdropping equipment, satellites and land-based antenna "farms" that intercept data transmitted between satellites and ground stations.'

> Despite the blanket secrecy the Soviet Union maintains about its military space program—secrecy that extends to denying such a program exists—the United States collects an enormous amount of information about Soviet space programs by rather simple means . . . even private individuals armed with basic radio gear have managed to decipher large portions of Soviet space operations. For years, the recently retired Geoffrey Perry, with the aid of his students at the Kettering Grammar School in England, has monitored the transmissions to and from Soviet satellites. With the aid of this data Perry has written . . . authoritative articles on the Soviet military space program, including Soviet photographic reconnaissance and signals intelligence satellites. It seems likely . . . that the Soviets can do at least as well deciphering the U.S. program as those schoolboys.

The most compelling evidence for Moscow that the United States was preparing to launch a new generation of SIGINT spy satellites, the *Post* argued, came not from browsing the Washington media but 'from observing U.S. intelligence collection activities in Alice Springs, Australia—an installation known as Pine Gap'.

As Robert Cooksey had found out, 'observing U.S. intelligence collection activities' could not easily be done in person. Anyone, however, could read the transcripts of parliamentary debates and ministerial statements in Hansard, which revealed that Pine Gap had been significantly expanded and upgraded in preparation for the launch of Orion-1.

Eighteen months earlier, on 8 July 1983, the Hawke government's first defence minister, Gordon Scholes, announced that an 'additional antenna' was to be built at the Joint Defence Space Research Facility at Pine Gap. The contract, valued at just under $1 million, had been awarded to Jennings Industries of Adelaide and covered the building of foundations for a new antenna pedestal and radome. The possibility of an extra antenna had been foreshadowed by the Fraser Coalition government in 1982, when it announced the installation of new computers and research equipment at Pine Gap.

On 15 November 1984, just over two months before the shuttle carrying Orion-1 was launched, the Australian defence department revealed that 'part of a new antenna' for Pine Gap would arrive in Alice Springs on 19 November 'in a US Air Force C5 Galaxy transport aircraft'. On 25 January, the day after the shuttle was launched, another statement captioned 'Equipment for Pine Gap' reported that 'a US Air Force C5 Galaxy will arrive at Alice Springs on Tuesday, January 29, carrying parts of the new antenna for the Joint Defence Space Research Facility ... A further flight is expected in April or May this year.' Senator Don Chipp, leader of the Australian Democrats, remarked three months later that 'the Galaxy C5 is the largest transport plane in the world. The volume of equipment being brought in would indicate that Pine Gap is again changing its function.'

Tipped off by such valuable snippets of information, the *Washington Post* said 'the Soviets can no doubt use their photographic reconnaissance satellites and, presumably, human agents, to observe an enormous amount of construction at Pine Gap ... Soviet analysts looking at the photographs from their satellites and then noting the increase in electronic signals in and out of Pine Gap could only conclude that Discovery had placed a new-generation SIGINT satellite into orbit.'

The desperate efforts by Weinberger and the US Air Force to withhold from the public any information about the Orion-1 reminded American journalists of the early days of the US satellite program. In those days the Pentagon was worried that publicity about US satellite flights over the Soviet Union would prompt the Kremlin to try to shoot them down, just as it had the U-2 spy plane flown by Gary Powers on 1 May 1960. It was not until 1978, nearly twenty years after the launch of the first Corona spy satellite, that the US government finally acknowledged the existence of its space-based photographic reconnaissance program. In the *Post*'s words, 'This high level of secrecy attached to US satellite reconnaissance activities is not justified on any grounds.'

In central Australia, the Galaxies kept coming.

Chapter 21
A magnet for protest

On 2 April 1985 three cyclists protesting against the Pine Gap base prevented a Galaxy from landing by riding onto the runway just as the giant transport prepared to touch down. It was the third time in six months that a Galaxy had arrived in Alice Springs with equipment for Pine Gap. According to a report in the *Canberra Times*:

> Airport and civil police, concentrating on 60 demonstrators waiting inside the airport fence, were taken by surprise. Police cars chased the cyclists for several hundred metres before catching them.
>
> When the runway was cleared, the Galaxy landed and unloading began of its huge cargo of alloy panels, believed to be for a new dome to house the base's latest antenna.
>
> While the Galaxy was being unloaded, three more demonstrators scaled the airport fence and ran onto the tarmac,

breaking through police lines and throwing orange paint over the aircraft before they were dragged away.

Pine Gap had been a magnet for protest since the late 1960s, when around 30 students from Adelaide's two universities announced their intention to hold a protest meeting outside the base. They also planned to distribute pamphlets in what one newspaper described as 'an attempt to force the people of Alice Springs to discuss the Pine Gap base'.

In 1974 a 'long march' from capital cities in the eastern states to the US base at North West Cape resulted in more than fifty arrests as well as highly publicised allegations of police brutality. Hailed by its organisers as an 'unqualified success', the North West Cape protest was described as 'but the beginning of a long campaign against US bases in Australia'.

Encouraged by the success of the North West Cape protest, the organisers announced a second 'long march' to Pine Gap. The base, they argued, played an 'integral role in the nuclear weapons delivery system of the US military machine'. If it could be 'put out of action', the South Pacific would be 'effectively denied to Yankee nuclear war plans'.

For protesters coming from Sydney the logistics for the long march to central Australia were daunting , although at $120 per person the cost was 'dirt cheap', according to an article in the University of New South Wales's student newspaper. Protesters would be travelling 6500 kilometres by bus; conditions would be 'very spartan' and would require passengers to live 'in a cooperative rather than a selfish way'. Cooking would be 'communal' and 'shared out amongst the participants'.

Under the heading 'WHAT WILL WE DO AT PINE GAP', the article said the nature of the demonstration would depend on our reception which is unpredictable. The group will have time to discuss plans on the way. We also intend to: strike the American flag, raise the Eureka flag and the Black Land Rights Flag, take possession of the land and return it to the Aborigines as a land rights gesture, dismantle the offensive nuclear equipment and conduct a tour of all prohibited zones ... Altogether the base covers 10 square miles of 'restricted areas'—hitherto its isolation has been sufficient to protect it from Australian wrath ... People are urgently needed to help organise for the trip, distribute leaflets, paste up posters, perform street theatre, and to assist in whatever way possible.

The station's geographical isolation could protect it from Soviet agents and eavesdroppers but not from domestic protesters, who converged on Pine Gap in ever-increasing numbers during the following decade.

In April 1981 more than 100 people marked the end of a seminar on US–Australian joint military facilities by demonstrating outside the base. When the protesters arrived, they found Northern Territory and Federal Police waiting for them outside the perimeter fence at the front gate of the base. Refused permission to deliver a petition to the chief of facility, Richard Krueger (who, in his twenties, had been sent by then CIA director Allen Welsh Dulles to recruit Einstein for the agency), they eventually handed it to the senior Federal Police officer on the scene. The protesters then walked a couple of kilometres around the perimeter to a ridge overlooking the base, shadowed the whole way by police, who watched but took no action. In fact the *Defence (Special Undertakings) Act 1952* allowed the arrest of anyone 'in the neighbourhood of' a prohibited area who was 'reasonably suspected of having committed, having attempted to commit or of being about to commit an offence against this Act'.

The aim of the Act was not only to protect against sabotage but also to maintain the wall of secrecy around the base. Section 9 stated that a person was guilty of an offence if he or she:

> (i) makes a photograph, sketch, plan, model, article, note or other document of, or relating to, an area or anything in an area; or
>
> (ii) obtains, collects, records, uses, has in his or her possession, publishes or communicates to some other person a photograph, sketch, plan, model, article, note or other document or information relating to, or used in, an area, or relating to anything in an area.

In the event, the protesters were allowed to depart the prohibited area unmolested, having done nothing more subversive than launching 'a large peace banner tied to a stack of helium-filled balloons'.

Two and a half years later a women's protest outside the Roxby Downs uranium mine announced its next stop would be Pine Gap. A column in the *Sydney Morning Herald* headlined 'Excursions' included both a 'women for survival' protest outside Pine Gap and a bike ride by 'men against Pine Gap'. Thirty Federal Police were rushed up to Alice Springs to deal with the demonstrators, who were said to be 'divided between women happy to make their protest on public land outside the base and those who want to take symbolic action—the "taking" of an American strategic base by Australian women'.

In November 1983 the Labor attorney-general, Gareth Evans, was forced to intervene after Northern Territory police broke up a demonstration outside Pine Gap and arrested more than a hundred protesters for breaching security at the base.

Allegations of women being stripsearched and forcibly fingerprinted and of 97 being held overnight in police cells prompted Evans to make a midnight telephone call to check on conditions of detention at the Alice Springs watchhouse. Asked for their names, all 97 women claimed to be 'Karen Silkwood'—the American nuclear laboratory technician who allegedly suffered contamination and died in mysterious circumstances in 1974.

Locals in Alice Springs were far from supportive of the protesters, with the mayor, Leslie Oldfield, declaring that the 'dole-supported, agitating, trouble-making' women-for-peace were not welcome in her town. In an editorial headed 'Ignore the peaceniks', the local newspaper said that local women did not consider either Pine Gap or the Americans as a problem.

By the time the women appeared before the magistrate, they had agreed to give their real names. Many were shocked, however, to be told that the maximum fine for trespassing on Commonwealth property was not $100, as they had thought, but $1000.

With several of the women 'resting' in Alice Springs after their night in the cells, others renewed their assault on the base. According to newspaper reports, territory police 'took a hard line' after the women ripped two sides of a gate from its metal supports and cut the wire fence around the perimeter. Despite their efforts to keep the demonstrators out of the base, around twenty managed to break in. The *Canberra Times* reported that helicopters 'buzzed overhead to radio the locations of those who managed to get inside the base and reached the thicket undergrowth. Several police officers in white overalls climbed down ropes from one helicopter hovering at about 15 metres when they spotted the women who had reached the bush.'

The seven women arrested during the first protest petitioned successfully to have their cases heard in Canberra. Eleven months later 50 protesters staged what the *Canberra Times* called a 'colourful' demonstration outside the ACT law courts as the women responded as a group to charges of trespassing and giving false names to the police. Appearing without counsel, they took it in turns to read from prepared statements. Their aim, they said, was not to argue their case on legal grounds but to make a political statement. 'We are here to proclaim our innocence and to condemn the legal system for protecting and legitimising a system of global terrorism and violence,' they said.

They accused the ACT court of complicity because it had 'nothing to say about the fact that the United States/CIA-controlled bases like Pine Gap unavoidably implicate Australia in the global nuclear war-machine'.

The magistrate fined the seven women $50 on each charge; six refused to pay and were jailed for two days, while the last defendant's fine was paid 'anonymously' after the hearing.

Unions, peace activists and the Labor left had never wanted the bases, and they wanted them even less now. Canberra's willing acceptance of US military bases contrasted sharply with the decision by New Zealand's newly elected Labour government to ban visits by US nuclear warships. Washington was outraged, but the move made Prime Minister David Lange a global hero of the anti-nuclear movement.

In March 1985 Lange attracted international attention by taking part in a debate at the Oxford Union in England, debating the proposition 'nuclear weapons are morally indefensible'.

The previous month his government had refused entry to an American guided missile destroyer, the USS *Buchanan*, because the Pentagon, in keeping with its standard policy, would neither confirm nor deny that the *Buchanan* had nuclear capability. Within days Washington cut intelligence and military ties with New Zealand, downgraded political and diplomatic exchanges, and withdrew its security guarantee to New Zealand (although the ANZUS Treaty remained intact).

New Zealand's anti-nuclear policy, and its willingness to stand up to what many saw as bullying by the United States, energised anti-US protesters in Australia. While Lange's government had turned back the USS *Buchanan*, Australia's Labor government not only welcomed visits by US nuclear warships but volunteered to help with the testing of the Pentagon's new MX intercontinental ballistic missile. The Hawke government offered to let US planes use Australian bases while monitoring the splashdown of MX missile warheads fired from California and aimed at the Tasman Sea.

The MX decision, made by the Cabinet's security committee without full Cabinet approval, split the Hawke government and attracted fierce public criticism. The indefatigable Des Ball fuelled the controversy with a research paper in which he named numerous Australian sites hosting portable satellite receivers that could be used to monitor and improve the accuracy of the MX missile's multiple warheads. As well as Canberra, Darwin, Perth, Townsville and Woomera, they included sites on Thursday Island and the Cocos Islands, Manus Island in Papua New Guinea and even Mawson and Casey stations in Antarctica. The US, reluctant to be caught up in a political brawl that would inevitably have brought more scrutiny on the joint facilities, decided to test MX without Australian help.

With twelve months left before the lease on Pine Gap expired, anti-American protests across the country continued. On 20 October 1986, a day of 'rain, hail and sleet', 800 women protested outside Parliament House in Canberra demanding the closure of Pine Gap. Hundreds then marched on the US Embassy. By the start of 1987 a broad coalition—including senators, unionists and entertainers such as the comedian Max Gillies and the singer Peter Garrett—had been forged with the aim of closing down all foreign bases, starting with Pine Gap. Protests went on all year, with more than 120,000 people marching through Sydney on Palm Sunday to demand the closure of Pine Gap.

The ten-year lease on Pine Gap formally expired at midnight on 18 October 1987. That day more than a hundred protesters, including the anti-nuclear senator Jo Vallentine, broke through the base's barbed wire outer perimeter fence. Police used solvent to free two demonstrators who had superglued themselves to a surveillance tower just inside the fence. The next day another 95 people were arrested. Some escaped arrest and were thought for a while to be at large inside the base, although the Australian defence representative at Pine Gap, Mike Busch, insisted that none had managed to get within half a kilometre of the inner perimeter. Vallentine would eventually serve three days in jail for refusing to pay a $250 fine for trespass.

In the end the nationwide protests against Pine Gap could not prevent the Hawke government from renewing the lease, although the terms of the new agreement were not publicly disclosed for more than a year. According to the terms of the previous extension, confirmed by an Exchange of Notes on 19 October 1977, after nine years either government could terminate the lease with one year's notice. Nurrungar operated under a similar agreement. On 22 November 1988 Bob Hawke told the parliament, 'We do not regard this as satisfactory, both because of the continuing importance of their effective operation to global peace, and because of the specific benefits to Australia of long-term access to their capabilities.' Instead, both governments agreed to extend the leases for Pine Gap and Nurrungar for another ten years, with three years' notice required to terminate either agreement.

Public pressure to close Pine Gap and other US facilities forced Hawke to go further than he had in his 1984 speech to justify their continued existence. He did this, above all, by emphasising their significance to international arms control. According to Hawke, without Pine Gap and Nurrungar 'the INF [intermediate-range nuclear force] treaty could not have been signed and the START [Strategic Arms Reduction Treaty] process would not have got under way'.

Nurrungar, Hawke said, was 'a ground station used for controlling satellites in the United States defence support program [DSP]'.

> The DSP satellites provide ballistic missile early warning and other information related to missile launches, surveillance and the detonation of nuclear weapons. Few if any elements of the strategic systems of either superpower make such a decisive and unambiguous contribution to keeping the peace as the defence support program... The DSP, through Nurrungar, would give the earliest warning of an international ballistic missile attack on the United States or its allies. Because the DSP gives longer warning of an attack than other systems, it reduces the chances that United States forces could be destroyed in a surprise attack,

and that makes it extremely unlikely that anyone would ever try such an attack. Together with other elements of the United States early warning system, the DSP provides highly reliable warning of attack. It thus plays a vital role in helping to prevent nuclear war by accident. Australians can be glad that we help to operate this vital facility, and that every day, around the clock, Australian personnel are . . . helping to prevent nuclear war.

Hawke was cagey about the function of Pine Gap, reiterating the long-established policy of all Australian governments not to comment on 'intelligence matters'. Since Pine Gap was—and always had been—'an espionage operation' (Des Ball's term), this appeared to prevent him from saying anything at all. While conceding there were 'limitations' on what he could state publicly about 'organisation and manning arrangements' at Pine Gap, Hawke's 1988 speech was notable both for what it revealed and for what it withheld.

At Pine Gap the number of Australians engaged in the central operational activity is being steadily increased. Some of these personnel, who are drawn from scientific and intelligence areas of the Department of Defence, are taking over functions previously carried out by United States employees. Whereas only a handful of Australian Government personnel was directly involved in the central work of the facility in the 1970s and early 1980s—contributing less than 10 per cent of the staff there—the proportion is scheduled to rise to about 30 per cent over the next two or three years. But Australians are not only doing more of the operational work at the facilities.

Under the new arrangements we have agreed, Australians will carry out more of the senior management functions at both Pine Gap and Nurrungar.

At Pine Gap, a senior Australian defence official will be appointed to a newly created position as deputy chief of the facility. He will advise and support the United States chief of the facility in managing the facility and its activities, and he will share responsibility with him for that work. He will also continue to be the officer in charge of the prohibited area, with ultimate responsibility for the physical security of the facility. Other senior management positions will also be filled by Australians.

Parallel changes have been agreed to the staffing at Nurrungar. Australians constitute some 40 per cent of the staff in the key operational areas, and will now take a role in management with the appointment of the senior Australian officer there as Deputy Commander. Like his counterpart at Pine Gap, he will share responsibility with the Commander for the management of the station and its physical security.

Many on the left remained unconvinced by Hawke's assertion of an equal partnership. The next day Jo Vallentine referred scathingly to the 'so-called joint facilities'. Another Nuclear Disarmament Party senator, Irina Dunn, moved that the Senate note 'with outrage' the extension of the lease and recognise that 'while Mr Hawke says that the bases are for verification purposes, it has been clearly established that they play an essential part in the USA first strike capability'.

Hawke's announcement that 'new' positions had been created for senior Australian officials to act as deputy chiefs of

Pine Gap and Nurrungar implied a harmonious and cooperative working relationship, but the truth was rather different.

While negotiations over the Pine Gap arrangement had gone relatively smoothly, the Australians ran into strong resistance at Nurrungar. A declassified report prepared by the US Air Force Space Command, entitled 'Background paper on Australian Implementing Arrangement—Deputy Commander controversy', reveals that Space Command and the Australian Department of Defence spent more than a year 'attempt[ing] to negotiate a new Implementing Arrangement . . . for USAF facilities in Australia' and that Space Command had worked with 'key OSD [Office of the Secretary of Defense] and Air Staff players to solve this impasse'. While it does not go into details of the 'controversy', the paper suggests that there was serious disagreement between the Americans and the Australians on the role of the deputy commander at Nurrungar as well as on other issues such as cost-sharing and security.

It was not until January 1990—fifteen months after Hawke told the Australian parliament that new staffing arrangements had been 'agreed to'—that Space Command considered all 'major' issues to have been 'resolved'. Although the Americans felt that the lengthy negotiations had yielded 'excellent results', the report ended with a recommendation for Space Command to 'retain negotiating authority to maintain the integrity of the IA [Implementing Arrangement] and all related instruments and issues'.

At Pine Gap another veteran CIA man, Donald Kingsley, had taken over as chief of facility. The Australian deputy chief of facility was John McCarthy, dispatched to Alice Springs from his previous job on the staff of Austalia's defence attaché in Washington. A newspaper profile coinciding with his appointment described McCarthy as 'the human face to the Hawke Government's resolution to demystify security and intelligence operations . . . the business-card spy, complete with non-silent telephone directory listing'.

Speaking more candidly than other Australians to have held senior positions at Pine Gap, McCarthy described the work of the base as 'the maintenance of world peace' and his own role as assisting Kingsley in the management of 'all aspects' of Pine Gap and being 'solely responsible for the facility's security'.

While emphasising the base's importance to international arms control verification, McCarthy pointed out that Australia had the right to 'fully utilise the equipment at Pine Gap' for its own purposes, including intelligence-gathering on 'the activities of Indonesia's defence forces and other Southeast Asian defence and intelligence operations'.

At the end of a decade of increasingly aggressive protest, Pine Gap was still in business. During those ten years the base had been significantly upgraded and expanded, while the spy satellites it controlled were bigger, more powerful and more intrusive than ever before. Rising public unease at the presence of the joint facilities was matched by the Australian government's growing assertiveness in determining how they were run and how they would be used. The new decade would begin with Australia preparing for a different kind of war in which America's geosynchronous satellites, controlled from Pine Gap and its sister stations, would be critical to the outcome.

Chapter 22
Desert Storm

The Hawke government's 1987 Defence White Paper stated that 'United States strategy is to deter war' and that 'Australia supports the concept of deterrence'. When the paper was published, the destruction of the Berlin Wall was still two years away and US military planners remained preoccupied with their Cold War enemy. According to the White Paper, the United States considered that deterring Moscow depended, in part, 'upon the credibility of the US capability for nuclear retaliation in the event of major Soviet attack upon the United States or its allies'.

The fall of the Berlin Wall on 9 November 1989 caused a permanent shift in the world's strategic balance. Between them, President Reagan and the Soviet leader, Mikhail Gorbachev, brought the Cold War to an end. Australia's joint facilities would remain central to both America's intelligence-gathering and war-fighting capabilities, but the enemy would change, and so would the role of Pine Gap.

A year earlier, the Joint Defence Space Research Facility had sloughed off a shifty euphemism ('space') and a blatant falsehood ('research') to become the Joint Defence Facility Pine Gap. In many ways, things went on as before: Pine Gap continued to be an indispensable collector of satellite-harvested technical intelligence for the CIA and the truth about its surveillance operations continued to be concealed from the Australian public. But in the years ahead the proportion of US military personnel at Pine Gap would increase, while the base itself became more directly involved in US military adventures around the world. From the early 1990s, US servicemen and women at Pine Gap would be heavily involved in controlling, processing and analysing signals intelligence downloaded from geosynchronous spy satellites.

David Rosenberg arrived at Pine Gap in October 1990, just as the military build-up was starting. As an intelligence analyst for the ultra-secret National Security Agency, Rosenberg was a civilian employee. His book, *Pine Gap*, although heavily redacted and short of technical detail, nevertheless offers valuable insights into the operations of Pine Gap during the eighteen years he worked there.

Rosenberg reveals that his first thought on seeing his new (mostly American and predominantly male) colleagues on the operations floor was 'How many Australians work here?' The question goes unanswered—the censor has blacked out the next five lines of type—but what follows is very interesting. Rosenberg writes that the military presence in operations began in 1990 (a paper for the Nautilus Institute dates it more precisely as September 1990) but 'wasn't visually obvious because Pine Gap had ... a No Military Uniform policy'. When the soldiers first arrived, he goes on, 'they were asked to "blend in" with the local community'.

Rosenberg began working at Pine Gap on 5 October 1990, two months after the Iraqi leader, Saddam Hussein, invaded and annexed neighbouring Kuwait. His arrival in central Australia coincided with the build-up of troops for Operation Desert Storm, a military operation by a coalition of 35 nations to expel Iraqi forces from Kuwait. By then 'preparations had already begun in earnest at Pine Gap to support the coalition against Iraq'. The base, he writes:

> was experiencing a long-term deployment of military personnel to supplement the previous 'civilian-only' Operations population... the United States Government was taking no chances in ensuring there were enough competent bodies to perform the work in Operations, and the newly arrived American and Australian military contingent (about five when I arrived) proved to be a capable, dedicated and useful resource that supplemented the civilians who historically occupied every billet within Operations.

Rosenberg is reticent about the intelligence role played by Pine Gap, although he states that the base was 'an essential part of the "silent" community that was in fact heard loud and clear at the highest echelons in government and by military leaders who were planning... Desert Storm'.

There is no doubt that the joint facilities at Pine Gap and Nurrungar supplied valuable information to coalition commanders both before and during Desert Storm, but their roles were different. Nurrungar, as Bob Hawke explained in his November 1988 speech to the House of Representatives, was the ground station that controlled satellites in the US defence support program (DSP). With their array of infrared sensors, the DSP satellites could detect the heat plumes of ballistic missiles and give early warning of missile launches within seconds of take-off.

Despite their inaccuracy, Saddam's Scud missiles were a major worry for the coalition, which feared they would be used to attack Israeli cities, dragging Israel into the conflict and threatening Arab support for the war against Saddam. Iraq used both a basic and an upgraded form of Scud. The DSP system had been designed to detect ICBMs; for a Scud to be detected, the infra-red intensity of the rocket exhaust had to be above a minimum level. Even the basic version was acknowledged after the war by the US Space Command to be at the edge of the detection capability of the DSP.

US intelligence cables confirm that US satellites kept a close watch on Iraq's Scuds during the weeks before coalition commanders launched Operation Desert Storm. A declassified but still heavily redacted memo prepared by the Iraq Regional Intelligence Task Force states that:

ON 2 DEC [1990] IRAQ LAUNCHED THREE SHORT-RANGE BALLISTIC MISSILES...FROM SITES IN EASTERN IRAQ...BOTH MISSILES DEMONSTRATED AN APPROXIMATE 600 KM FLIGHT AND A SEVEN-MINUTE TIME OF FLIGHT. THE IMPACT OF BOTH MISSILES OCCURRED...IN THE VICINITY OF THE WESTERN SCUD LAUNCH COMPLEX OF WADI AMIJ, IRAQ. THE SECOND LAUNCH WAS A POSSIBLE MISSILE FAILURE...IT IS POSSIBLE THAT THIS MISSILE ACTIVITY REPRESENTS OPERATIONAL TESTING OF IRAQ'S INDIGENOUSLY-PRODUCED MOBILE SCUD TEL [LAUNCH VEHICLES]...THESE TEST FIRINGS MAY BE A SIGNAL IN RESPONSE TO THE RECENT US FORCES ALERT.

The Scud report drew on data obtained from DSP launch-detection satellites—the satellites Bob Hawke had spoken about in his November 1988 statement to the House of Representatives. The DSP satellites identified ballistic missile launches and collected data including missile type, launch site and impact area.

A declassified memorandum entitled 'Operations Desert Shield and Desert Storm Assessment', prepared in January 1992 by the US Space Command, confirms the importance of the Nurrungar station as well as hinting at the politics involved. It reveals that as early as 8 September 1990, more than four months before the air war began, 'DSP ground sites were enhanced to provide better SCUD warning... (Note. Negotiations were conducted between the Government of Australia and USSPACECOM to support GOA requirements for notification to the Minister of Defense upon implementation of these procedures at Woomera, Australia. The DSP large processing station at Woomera is jointly manned with Australians. Operation of the site is subject to United States and Australian agreement. New procedures were established for the SPACC to notify the GA [GOA] when a specific SCUD warning operation was implemented.)'

In late December the *Los Angeles Times* reported that Iraqi test firings described in the cable had enabled the US to make important adjustments to its launch-detection system. Two more Scud launches took place in the last few days of December, the latter causing a 90-minute alert of allied forces.

According to a report published twelve months later in *Aviation Week & Space Technology*:

Iraqi President Saddam Hussein 'blundered' when he launched... test Scuds before the war began. The tests enabled the US to ascertain the improvements needed to lengthen the warning time of a Scud attack.

In his paper 'The Intelligence war in the Gulf', Des Ball writes that an advanced DSP satellite launched in December 1990 was able to broadcast intelligence about Scud launches directly to mobile receiving stations in Saudi Arabia. Timely information from DSP satellites made it possible to find and perhaps destroy mobile launch vehicles.

As well as detecting Scud launches, the DSP system was also able to make predictions about a missile's likely impact point. The success of US Patriot missiles in shooting down incoming Scuds was widely attributed to the information supplied by DSP satellites. Ball's Gulf War paper quotes the director of the Strategic Defense Initiative Organization, Henry Cooper, saying, 'This was the first war in which space played a central part, and DSP was a very important part of it.'

A declassified 1991 memorandum entitled 'DSP Desert Storm summary briefing', prepared by the US Air Force Space Systems Division, concludes that DSP 'significantly contributed to US warfighting capability'. It highlights the success of the DSP in two key areas:

- Scud early warning
- Scud targeting.

The January 1992 memorandum 'Operations Desert Shield and Desert Storm Assessment' states:

In addition to warning, the elimination of mobile-SCUD launchers was a top priority and one of the most difficult tasks of the war. USSPACECOM provided launch locations identified by DSP, allowing US Central Command... to vector strike aircraft to attack mobile-SCUD launchers.'

Nurrungar was one of several Australian bases supplying intelligence to coalition forces in the Gulf. By the time the Hawke government committed Australian naval ships to the Middle East, other joint facilities were already providing valuable support to US military operations. Senator Paul McLean told the Australian Senate, 'The United States-Australian bases at Pine Gap, Nurrungar and North West Cape... supplied vital intelligence, surveillance and communications to the Pentagon from the outset.' According to Nuclear Disarmament Party senator Jo Vallentine, intelligence—rather than a 'token' naval deployment—was Australia's major contribution to the Gulf War. Pine Gap and Nurrungar, she told the Senate, were 'feeding satellite pictures, intercepting military messages, and feeding information to the cashbox coalition so that it can sometimes accurately get its targeting right'.

When the coalition launched Operation Desert Storm on 17 January 1991, the US had half a dozen geosynchronous spy satellites sitting over the western Indian Ocean to eavesdrop on Iraqi communications and other electronic emissions. The satellites were controlled from Pine Gap and from ground stations in Britain and Germany.

A key target was telephone communications between Saddam and his generals. Intelligence from intercepted Iraqi communications was conveyed almost instantaneously not just to CIA headquarters and the Pentagon but to US generals in the Gulf. During the build-up of coalition forces that preceded Desert Storm, *US News & World Report* noted that US spy satellites originally designed to intercept Soviet signals were 'eavesdropping on Iraqi communications and constantly updating the picture of the region available to American field commanders'.

Intercepting Iraqi communications was one thing; analysing and, if necessary, deciphering them was another. The Iraqis, who had just fought an eight-year war against Iran, were disciplined operators. Ball has noted that the National Security Agency, the US spy agency with primary responsibility for signals intelligence, and its sister organisation, the Service Cryptologic Agencies, were initially frustrated by the Iraqi investment in 'sophisticated Soviet, British and French encryption systems; the widespread use of landlines and optical fibre cables; the high standard of discipline exercised by Iraqi radio operators; and a scarcity of Iraqi linguists in the US intelligence community'. Furthermore, the US had supported Saddam's armed forces with high-grade intelligence from its satellites during the Iran–Iraq war. This had given the Iraqis a fair idea of the capabilities of US spy satellites and they were able to take countermeasures to hamper US surveillance.

After the launch of Desert Storm, decrypting of Iraqi communications quickly became easier. In part this was due to the sheer volume of data being collected once the shooting war began, not only by geosynchronous satellites but by airborne and ground-based SIGINT systems as well. The more intelligence that poured into Pine Gap and other ground stations, the better the chances of breaking Iraqi codes. The fact that secure Iraqi communications systems had been destroyed by the bombing campaign that preceded the ground war also forced Iraqi commanders to resort to less secure systems. Communication links between the high command in Baghdad and commanders in the field began to collapse. Within a month of the air war starting, so much had been destroyed that, according to media reports, Saddam was having to send messages using dispatch riders on motorbikes.

The objective was not the total destruction of Iraqi communications; rather, it was to wipe out enough of the military's secure communications network to force the Iraqi commanders to discuss military plans and preparations using easily intercepted high-frequency radios. Simply monitoring these calls enabled the coalition to pinpoint the physical location of top Iraqi leaders, including Saddam himself, and on occasions to target them with air strikes.

Ever since Australia's defence minister, Allen Fairhall, assured reporters in 1969 that Nurrungar had 'no offensive capacity', successive ministers and prime ministers (when they spoke about them at all) had stressed the 'defensive' function of the joint facilities and played up their role in international arms control. In November 1988 Bob Hawke had reiterated the base's role in the 'verification of arms control and disarmament agreements'. On 18 January 1991, the day after Desert Storm began, Hawke was evasive. Asked about the involvement of the joint facilities in detecting Iraqi missile launches, the prime minister told reporters 'They're not irrelevant, let me put it that way.'

The US president and commander in chief, George Bush, was more candid about the military value of signals intelligence. In May 1991 Bush visited the National Security Agency headquarters at Fort Meade, Maryland to pay tribute to the 'unsung heroes' whose secret surveillance work helped win the war.

Chapter 23
The canteen tour

Throughout the 1990s Pine Gap continued to expand. In August 1994 Labor's then defence minister, Robert Ray, informed the Senate that there were 725 staff at Pine Gap, of whom 405 were Australian. He revealed that the site now had ten radomes and associated antennae, and three additional satellite antennae. The expansion accelerated with the election of a Coalition government in March 1996. The following year the new minister for defence, Ian McLachlan, announced a major staff increase at Pine Gap involving both 'civilian and military personnel'.

Meanwhile bases in other parts of Australia had disappeared. Many of the minor facilities listed by Allen Fairhall in 1967—such as the Cooby Creek Tracking Station near Toowoomba in Queensland, and the satellite tracking and data acquisition network facility at Orroral Valley in the ACT—had been closed by the end of the 1980s. Of the major bases, the US naval communications station at North West Cape had been wound back since the retirement of the navy's Polaris nuclear submarines and was dispensed with altogether in the early 1990s, while the ageing ballistic missile early warning station at Nurrungar was decommissioned in 1999, its responsibilities transferred to Pine Gap. (The next millennium would see a renewed role for North West Cape in US space surveillance and military signals intelligence, but Nurrungar would remain derelict.)

Pine Gap alone had survived three decades of political controversy and public protest to become an indispensable link in America's global intelligence network. The ten-year lease negotiated by the Labor government came up for renewal in November 1998. Since the previous extension a special parliamentary committee—the Joint Standing Committee on Treaties—had been created to determine whether international treaties were or were not in Australia's national interest. It was, in the words of one member, a 'fair dinkum committee', not a rubber stamp. Report 26 dealt with the agreement between Australia and the United States to extend the lease on Pine Gap.

The Australian political landscape had changed, but the issue of the joint facilities was proving as divisive under John Howard as it had been under Bob Hawke. While access to Pine Gap was less restricted than it had been in the days of Holt and Gorton, Australia's Department of Defence had no intention of throwing open the gates to a parliamentary committee.

Frustrated in its efforts to obtain information about the facility, the joint committee turned to the most knowledgeable non-government authority: the Australian National University's Professor Des Ball.

The decision to call Ball—who in the 1960s had been identified by ASIO as a 'person of interest'—as a witness in open hearings looked like a calculated rebuff to the Department of Defence. Ball, after all, had built a distinguished academic career on finding out and publishing information about Pine Gap and other US bases that the department would have preferred to remain secret.

The committee chair, Andrew Thomson, would later admit his colleagues' frustration at the defence department's refusal to cooperate, noting that the department had provided the committee with 'less information about the facility than is available at the public libraries'.

There was clearly a conflict, Thomson told the parliament, between 'the openness and the transparency that characterises the government's reformed treaty-making process and the Department of Defence's tradition of secrecy'.

Whatever ASIO thought of him, Des Ball was not a loose talker. He knew much more about the world of intelligence than he was prepared to reveal publicly. While there was a 'large arena of signals intelligence activities' that Ball considered it 'quite proper to talk about', he drew the line at revealing 'technical operational secrets' and the 'sensitive intelligence collected through that technology'. His reasons were professional as well as ethical: Ball's access to defence and intelligence officials depended on his discretion.

It was Ball, however, who reminded the joint committee that behind the 'public agreement' they were trying to assess lay a 'classified agreement'. It was the classified agreement, he said, that 'sets out the command and control arrangements ... the criteria for scheduling what satellites are going to listen to what emissions ... the arrangements for dissemination of the intelligence ... [and] the conditions for security, not just the physical security of the Pine Gap facility but also security of the intelligence which is collected at Pine Gap. It sets out who in Australia is allowed to see that, and who at the various levels is allowed to visit the facility ... It sets out what level of intelligence people are allowed to see, whether they are only allowed to see finished assessments or whether they are allowed to see actual raw intercepts.'

There was a hint of mischief in Ball's suggestion, at the end of his prepared statement, that it was 'worthwhile this committee spending some time trying to learn what it can from the Department of Defence ... about the classified agreement'.

The committee, as Ball surely knew, had as much chance of being allowed by the defence department to study the details of the classified agreement as of being invited by the CIA to peruse the codes in the National Cryptographic Room.

As far as the committee was concerned, the lack of co-operation by the Department of Defence 'raise[d] a fundamental issue about accountability and responsible government'. For all the apparent 'openness' of Bob Hawke's 1988 statement about Pine Gap, the truth was that a decade later the Australian parliament had (in the words of Senator Helen Noonan) 'nil' oversight of its rules of operation. Exasperated by the withholding of crucial information, the committee was reduced to begging its star witness for 'any constructive suggestions about how we might better inform ourselves about the ... classified agreement'.

Ball's response revealed something about his own research methods and about the source of his remarkable knowledge:

> It really depends on the relationship which this committee has with people in the Department of Defence, the personal levels of connections which have been built up and the trust in the end which exists between members of this committee and the Department of Defence as to the extent to which they might be a little bit more forthcoming with you.

In his prepared statement Ball outlined how Pine Gap had evolved from a ground station listening to Soviet missile telemetry signals to a station capable of intercepting everything from military radar to private cellphone conversations. Satellites controlled from Pine Gap were positioned close enough to intercept signals sent from the ground to communication satellites parked in geostationary orbits. They also monitored a wide range of microwave emissions on the Earth's surface, including long-distance telephone calls transmitted via terrestrial circuits. As well as eavesdropping on military communications, the satellites were capable of intercepting channels used by political and government agencies—and even private citizens.

It would be another fifteen years before the full ramifications of what Ball told the joint committee in August 1999 were laid bare by a National Security Agency whistleblower, Edward Snowden.

Ball was a longtime critic of other US bases in Australia, such as North West Cape, which had been set up in 1963 to communicate with US submarines, and Nurrungar, but when it came to Pine Gap, he admitted to having 'force[d] myself to come out in support' because he considered 'the intelligence which is collected there as critically important and collectable in no other way'. He also acknowledged that safeguards existed to ensure that 'if anything is intercepted relating to economic intelligence or political intelligence or personal communications of Australian citizens, that does not get any further than the Australian representatives on the ground at Pine Gap'.

Des Ball was not the only witness the joint committee called that day. Another was Professor Paul Dibb, a former defence department insider and author of the 1987 Defence White Paper. In his paper 'The "Joint Facilities" revisited: Desmond

Ball, democratic debate on security, and the human interest', Richard Tanter remarks that:

> Dibb followed [Ball], making clear the limitations on what he could tell the Committee. The sequencing was intentional and significant, with Dibb listening to Ball's presentation. Not only did Dibb not dissent from any of Ball's analysis, but he went on to give it his overt, if partial, imprimatur: 'Professor Ball can go somewhat further than I can. He and I would have significant areas of agreement.'

One area of agreement was that the committee members would learn little by visiting Pine Gap, although visiting Pine Gap was what the committee seemed to want most. 'We cannot get through the gate, and it is on our land,' Labor senator Chris Shacht protested. 'It is in Australian sovereign territory. Members of a foreign power can get through the gate; we cannot.'

Getting through the gates of Pine Gap had been the ambition of nosy Labor parliamentarians since the late 1960s, when defence minister Allen Fairhall had made it his business to keep them out. The rules governing access had been relaxed since then, but Pine Gap's operating secrets remained well hidden. Visitors were shown hospitality but not much more. Ball told the committee:

> You would walk in the outer security gate or drive to the outer security gate. You would then go into the internal one. They would take you in and give you donuts and coffee. You would see . . . about 18 satellite antennas sitting around the facility. You would see an enormously large computer room with a lot of guys sitting there with earphones and other

things. But it is not going to tell you very much about what goes on there.

Dibb was equally dismissive: 'If you talk about access, let me put it to you very clearly: the access you would not want is the access where you would go through the front door and you would see the domes, the generator . . . and the canteen.' Unsatisfactory as it was, even this was more access than the joint committee had been given. Senator Schacht complained that he and his fellow committee members 'cannot even get the canteen tour'.

In the end the joint committee decided unanimously—but without much conviction—that keeping Pine Gap was in Australia's national interest. Tabling the committee's report on 18 October 1999, Andrew Thomson drew attention to its final words:

> . . . with the limited evidence made available, we find no reason to object to the continuation of the Joint Defence Facility.

It was hard to see the verdict as anything other than a foregone conclusion. The joint committee's investigative diligence had been no match for what Thomson described as the Department of Defence's 'tradition of secrecy'.

But plans for Pine Gap went far beyond 'continuation'. By 2000 the workforce was expected to be 895 (420 Australians and 475 US personnel)—an increase of nearly 25 per cent in just six years. More new radomes were appearing: by 2002 there would be fourteen, plus another twelve satellite antennae.

In the 1970s the main intercept antenna on geosynchronous satellites controlled from Pine Gap had been around twenty metres in diameter. By the start of the new millennium the intercept antenna on the most modern satellites was roughly 100 metres in diameter—big enough to be seen from Earth with a small telescope. Big enough to listen to terrorists chatting on their mobile phones.

Chapter 24
Hunting for Osama

On 11 September 2001 terrorists hijacked four passenger planes and flew two of them into the twin towers of New York's World Trade Center and a third into the Pentagon. The last plane crashed in Pennsylvania after passengers tried to wrestle control from the hijackers. Nearly three thousand people died in New York, Virginia and Pennsylvania, as well as the nineteen hijackers. Intercepted communications quickly pointed to the al-Qaeda chief Osama bin Laden as the leader of the plotters.

It is possible that intelligence capable of warning US authorities about the 9/11 attack was intercepted by satellites controlled by the National Security Agency from Pine Gap but not analysed in time to provide a warning.

In his book *Body of Secrets*, updated a few months after 9/11, the US author James Bamford revealed that as 'tens of millions of communications continue to be vacuumed up by NSA every hour, the system has become overwhelmed as a result of too few analysts'. John Millis, an ex-CIA officer and former staff director of the House Select Committee on Intelligence, told Bamford, 'We don't come near to processing, analyzing, and disseminating the intelligence we collect ... We're totally out of balance.'

According to the former NSA director Michael Hayden, 'NSA is in great peril ... In the previous world order our primary adversary was the Soviet Union. Technologically we had to keep pace with an oligarchic, resource-poor, technologically inferior, overbureaucratized, slow-moving nation-state. Our adversary communications are now based upon the developmental cycle of a global industry that is literally moving at the speed of light ... cell phones, encryption, fibre-optic communications, digital communications.' Numbers told the story:

> Forty years ago there were five thousand standalone computers, no fax machines, and not one cellular phone. Today, there are over one hundred eighty million computers—most of them networked. There are roughly fourteen million fax machines and forty million cell phones ... The telecommunications industry is making a one trillion-dollar investment to encircle the world in millions of miles of high bandwidth fiber-optic cable. They are aggressively investing in the future ... Osama bin Laden has at his disposal the wealth of a three trillion dollars-a-year telecommunications industry.

Millis, the ex-CIA man, put it more succinctly: 'Signals intelligence is in a crisis ... in the last four or five years technology has moved from being the friend to being the enemy.'

Although it was unable to prevent the catastrophe of 9/11, the NSA threw itself into the task of finding Osama bin Laden and other members of al-Qaeda involved in planning the attack. As new scraps of information about the terrorists were uncovered—such as the names of the hijackers obtained from passenger manifest lists—critical intelligence or 'CRITIC' messages were flashed across secure NSA lines to field stations around the world.

Until 1998 bin Laden had used a satellite phone to communicate with his lieutenants from his base in Afghanistan. According to Bamford, the phone was used for 'hundreds of calls' to London, Iran, Saudi Arabia, Pakistan and Sudan. Some of the calls to and from bin Laden were intercepted by the Australian Defence Signals Division's listening post at Geraldton in Western Australia before being encrypted and forwarded to the NSA. The NSA intercepted his calls and agency officials had been known to play recordings of bin Laden talking to his mother in Syria. Bin Laden stopped using the satellite phone in 1998 after President Clinton ordered a cruise missile attack on his camp in retaliation for the bombing of two US embassies in East Africa.

Although it could no longer eavesdrop on bin Laden himself, the NSA was still able to pick up the odd call between his al-Qaeda associates. David Rosenberg recalls staff at Pine Gap joining in the task of 'hunt[ing] down those behind the attacks' and 'diligently assessing Afghanistan's weapon systems and communications networks' as a US-led coalition prepared for the bombing campaign that would precede the invasion of Afghanistan.

Both Pine Gap and Geraldton belong to a global system of satellite communications monitoring known as Echelon, a key component of the Five Eyes intelligence-sharing network linking the United States, Australia, Britain, Canada and New Zealand. In his paper 'Australia and the "Five Eyes" intelligence network:

The perils of an asymmetric alliance', Andrew O'Neil writes that Echelon has been operating since the 1960s and involves 'the interception from civilian satellites of communications on the basis of keywords submitted by each Five Eyes member country. Through highly sophisticated computer programs (called "dictionaries") at ground relay stations, these communication intercepts are sorted according to these keywords and... automatically forwarded to analysts.'

Powered by the almost limitless harvesting and storage capacity of the NSA (equal to '5 trillion pages of text—a stack of paper 150 miles high', according to James Bamford), Echelon was fundamentally different to other Cold War–era SIGINT interception systems. Originally designed to eavesdrop on diplomatic and miltary traffic, Echelon had a broadband capacity that enabled it to monitor virtually all electronic communications between public and private organisations and ordinary citizens throughout the world.

A 194-page report published by the European parliament on 11 July 2001 described Echelon as 'a global system for the interception of private and commercial communications'. Echelon, it said, had the capacity to carry out 'quasi-total surveillance. Satellite receiver stations and spy satellites in particular are alleged to give it the ability to intercept any telephone, fax, Internet or e-mail message sent by any individual and thus to inspect its contents'.

New Zealand journalist Nicky Hager told the Europeans that the only countries not being spied on were the Five Eyes member states themselves.

Like Des Ball in Australia, Hager had a long history of obtaining information the government wanted to remain secret. In his introduction to Hager's book *Secret Power:*

New Zealand's role in the international spy network, David Lange notes that Hager had 'long been a pain in the establishment's neck' and that 'an astonishing number of people have told him things that I, as Prime Minister in charge of the intelligence services, was never told'.

Hager gave a detailed description of Echelon in his book, which was published five years before the European parliament's report.

> Each station in the ECHELON network has computers that automatically search through the millions of intercepted messages for ones containing pre-programmed keywords or fax, telex and e-mail addresses. For the frequencies and channels selected at a station, every word of every message is automatically searched...
>
> Computers that can search for keywords have existed since at least the 1970s, but the ECHELON system has been designed to interconnect all these computers and allow the stations to function as components of an integrated whole...
>
> Under the ECHELON system, a particular station's Dictionary computer contains not only its parent agency's chosen keywords, but also a list for each of the other four agencies. For example, the Waihopai computer [in New Zealand] has separate search lists for the NSA [USA], GCHQ [UK], DSD [Australia] and CSE [Canada] in addition to its own. So each station collects all the telephone calls, faxes, telexes, Internet messages and other electronic communications that its computers have been pre-programmed to select... and automatically sends this intelligence to them. This means that the New Zealand

stations are used by the overseas agencies for their automatic collecting—while New Zealand does not even know what is being intercepted from the New Zealand sites for the allies.

Some of Hager's comments about the New Zealand stations were also true of the listening posts in Australia. Duncan Campbell, the British journalist whose revelations about Echelon prompted the European parliament's investigation, reported in the Melbourne *Age* that 'about 80 per cent' of the messages intercepted at the Geraldton station were 'sent automatically from its Dictionary computer to the CIA or the NSA, without ever being seen or read in Australia'. Unlike many others, however, Campbell did not believe that Echelon could intercept phone calls, asserting in his July 2000 paper 'Inside Echelon' that:

> Neither Echelon nor the electronic espionage system of which it is a part are capable of doing so. Nor does the equipment have the capacity to process and recognise the contents of each voice message or telephone call.

Duncan Campbell's work was cited numerous times in the European parliament's report, especially his allegations that Echelon had been diverted from its original purpose of spying on the Soviets and was now being used by the United States for industrial espionage.

His and Hager's allegations about Australian involvement in Echelon had been corroborated, however, by an unlikely source: the head of the Defence Signals Directorate (DSD), Martin Brady.

The DSD was the main Australian defence intelligence agency actively involved in operations at Pine Gap and was reported to have had a presence at the base since as early as 1977. By 1980, according to a 2016 report by the Nautilus Institute, the DSD was analysing SIGINT collected at Pine Gap, to which it had direct access. By the mid-1980s material intercepted at Pine Gap was being studied at DSD headquarters in Melbourne's Victoria Barracks alongside SIGINT picked up at the DSD's own listening posts. After the DSD shifted its headquarters to Canberra in the early 1990s, a 'substantial' part of the complex was devoted to analysing intelligence from 'the Desert'.

In March 1999 Brady replied to a series of questions from Ross Coulthart, a journalist on Channel Nine's *Sunday* program. Without naming Echelon, the DSD chief confirmed that 'DSD does cooperate with counterpart signals intelligence organisations overseas under the UKUSA [Five Eyes] relationship. Both DSD and its counterparts operate internal procedures to satisfy themselves that their national interests and policies are respected by others. In Australia's case, these processes are subject to review by the Inspector-General.' He insisted that the rules 'prohibit the deliberate interception of communications between Australians in Australia; the dissemination of information relating to Australian persons gained accidentally during the course of routine collection of foreign communications; or the reporting or recording of the names of Australian persons mentioned in foreign communications'.

While Coulthart's questions related primarily to Echelon, Brady opened a window on other Australian-based surveillance programs with his statement that the DSD 'directly contributes to the military effectiveness of the ADF [Australian Defence Force]'.

Not surprisingly, Brady offered no evidence for his assertion that intelligence collected in Australia supported military operations, but former Pine Gap insider David Rosenberg supplied several examples in his book. One of these involved the monitoring of Iraqi radar emissions during Desert Storm. According to Rosenberg, the Iraqis used high-altitude balloons to assess atmospheric conditions before the launch of a Scud missile. These balloons were tracked using a special radar unit known as 'End Tray'. Rosenberg writes that End Tray was a 'fairly common signal' for the operators at Pine Gap to locate. In theory, picking up an End Tray emission enabled allied planes to find and destroy the mobile launcher, but there was little evidence of these radar intercepts leading to the destruction of even a single mobile Scud launcher. 'This result has been studied,' Rosenberg writes, 'and the United States has undoubtedly prepared a "Lessons Learned" report that should enable a more effective response against mobile ballistic missiles in the future.'

Responsibility for the failure to fully exploit the US-led coalition's vast resources in space lay primarily with the National Reconnaissance Office (NRO), which had been established in 1961 to manage America's spy satellite program (its existence was not acknowledged by the US government until 1992). The NRO played a key role at Pine Gap. According to a 1984 paper by Des Ball entitled 'US installations in Australia—agenda for the future', while Pine Gap was 'operationally controlled' by the CIA, it was 'formally administered' by the NRO. Spy satellites controlled from Pine Gap, which had originally been used to verify arms control treaties, were being enhanced and reoriented towards warfighting, but the first Gulf War exposed serious deficiencies.

General Norman Schwarzkopf, commander of the coalition forces, complained about the speed and accuracy of satellite intelligence, telling Congress after the war that the intelligence community 'should be asked to come up with a system that will . . . be capable of delivering a real-time product to a theater commander'. The House Armed Services Committee later criticised satellite imagery for having been 'often late, unsatisfactory, or unusable'. Presidential Decision Directive 35, issued by President Clinton in March 1995, identified support for military forces as a top intelligence priority.

At a congressional hearing in March 1997 the deputy director of the NRO, Keith Hall, admitted that while the NRO had 'cobbled together' a workable space surveillance infrastructure during Desert Storm, it was 'obvious that direct space support to the warfighter was immature at best'.

The following year Hall told Congress that in future US forces would 'rely upon space systems for global awareness of threats, swift orchestration of military operations, and precision use of smart weapons . . . Our goal is to detect, track and target anything of significance worldwide and to get the right information to the right people at the right time.'

Pine Gap was critical to this mission. In his book David Rosenberg writes that during the war in Kosovo he and his colleagues at Pine Gap 'found, analysed and relayed information' in support of the coalition forces invading Kuwait and Iraq, and that in 1998 and 1999, during the war in Kosovo, its satellites 'monitored various Serbian military-related signals' as well as 'providing SIGINT to tactical planners and assisting during the search-and-rescue phase of the conflict'.

By the time coalition forces invaded Afghanistan in October 2001, the quality and availability of satellite intelligence had dramatically improved. According to Carl Bloggs's book *Masters of War: Militarism and blowback in the era of American empire*, new information-distribution technology enabled intelligence to be sent not just to ships and fighter jets but to 'individual special operations soldiers'. Using a relay station on the island of Diego Garcia, video data from Predator drones was 'fed . . . directly to special operations soldiers on horseback in the Afghan countryside'.

After the invasion, Pine Gap played a key role in the Pentagon's controversial program of assassinations by armed drones. Former Pine Gap staff told Fairfax Media's Philip Dorling in 2013 that the base had had 'outstanding' success in coordinating drone strikes in Afghanistan and Pakistan. Between 2500 and 3500 al-Qaeda and Taliban militants were reported to have been killed in targeted assassinations. 'The [Taliban] know we're listening, but they still have to use radios and phones to conduct their operations, they can't avoid that', a former operator told Dorling. 'We track them, we combine the signals intelligence with imagery, and once we've passed the geolocation intel[ligence] on, our job is done. When drones do their job we don't need to track that target any more.'

According to 'Australian Defence intelligence sources', data collected and processed at Pine Gap had undergone a 'massive quantitative and qualitative transformation' during the previous decade and 'especially the past three years'. One unnamed official told Dorling that the US would 'never fight another war in the eastern hemisphere without the direct involvement of Pine Gap'.

Among the activities listed by National Security Agency officers who had recently served at Pine Gap were 'signals intelligence collection, geolocation . . . and reporting of high-priority target signals' including 'real-time tracking'.

Pine Gap was also involved in the 'red dot' system that

warned troops about radio-controlled improvised explosive devices, or IEDs. The director of the NRO, Bruce Carlson, told *Time* magazine in 2011 that soldiers travelling in humvees along Afghan roads saw red dots on their computer screens alerting them to the location of suspected IEDs. While the system showed a lot of 'false positives', most were weeded out before the alerts were sent. He estimated the 'red dot' program was '80 per cent effective'.

Since the 1970s, when more than 800 jobs were cut from the CIA's espionage division, there had been concerns within US intelligence that technological advances were coming at the cost of agents on the ground. President Carter's director of the CIA, Stansfield Turner, made no secret of his preference for 'technical collection' (using satellite imagery and data interception) over the recruitment of 'moles' by CIA officers. Critics who argued that the downgrading of 'human intelligence' was making it harder for America to anticipate aggressive acts by its adversaries were proved right when the CIA failed to predict either the overthrow of the shah of Iran or the Soviet invasion of Afghanistan.

The technological trend continued and even accelerated over the following decades. In August 2001, three years after al-Qaeda bombed two US embassies and just a few weeks before 9/11, a disgruntled former CIA officer, Reuel Marc Gerecht, wrote an article in the *Atlantic Monthly* in which he described America's counterterrorism program in the Middle East as a 'myth'. The CIA, he wrote, had done little to cultivate undercover operatives who could blend in with the Islamic radicals it was supposed to be watching. Behind-the-lines counterterrorism operations were deemed 'just too dangerous for CIA officers to participate in directly'. A former 'senior Near East Division operative' told him 'the CIA probably doesn't have a single truly qualified

Arabic-speaking officer of Middle Eastern background who can play a believable Muslim fundamentalist who would volunteer to spend years of his life with shitty food and no women in the mountains of Afghanistan. For Christ's sake, most case officers live in the suburbs of Virginia.'

Gerecht concluded, 'Unless one of bin Laden's foot soldiers walks through the door of a U.S. consulate or embassy, the odds that a CIA counter-terrorist officer will ever see one are extremely poor.'

The warnings by Gerecht and others—about the risks of relying on technology and about the diminution of 'human intelligence'—proved to be prescient. Pine Gap and Echelon had been unable to prevent the 9/11 attacks and were little help in finding bin Laden, who did not use either phones or the internet. US intelligence spent another ten years hunting for the al-Qaeda mastermind. In the end it was not satellite intercepts that led the CIA to his hideaway in the Pakistani town of Abbottabad, but a careless al-Qaeda courier. Bin Laden was killed on 2 May 2011 during a raid on his compound by US special forces.

Chapter 25
Snowden

In the early hours of Friday, 9 December 2005, four peace activists protesting against Australia's involment in the Iraq war broke into the Joint Defence Facility Pine Gap. Jim Dowling, Adele Goldie, Donna Mulhearn and Bryan Law were members of an organisation calling itself Christians Against ALL Terrorism. The four, clad in white overalls with insignia identifying them as a 'Citizens Inspection Team', had marched across several kilometres of spinifex scrub and used bolt cutters to cut through both the outer and inner perimeter fences. The deputy chief of the facility would assert at their trial that the 'Pine Gap 4' were the first activists ever to have breached the 'technical area' housing the radomes and the supercomputers.

The four were quickly arrested and charged with intentionally causing damage to property (a charge that carried a maximum penalty of ten years' gaol) and with having breached two clauses of the *Defence (Special Undertakings) Act 1952* by entering and using a camera within a prohibited area. Passed before the British nuclear weapons tests in the Montebello

Islands off Western Australia, the Act was designed to keep the public out of sensitive military installations and had never been used before. The maximum penalty for being found in a prohibited area was seven years' imprisonment.

That the protesters made it as far as they did was, they believed, something of a miracle. The head of security at Pine Gap, Inspector Ken Napier of the Australian Federal Police, admitted to having told them just hours before the break-in that they had no chance of penetrating the 'technical area'. But penetrate it they did, and at their jury trial in the Northern Territory Supreme Court the Crown demanded they be convicted and gaoled. All four were found guilty but Justice Sally Thomas rejected the prosecution's call for imprisonment, sentencing them to modest fines of between $500 and $1250 and ordering them to pay $10,075.89 for damage to the fences. The Crown appealed on the ground that the lenient sentences were 'manifestly inadequate' and 'shocked the public conscience', while two of the accused cross-appealed, alleging errors by the trial judge.

While the Commonwealth was not a party to the prosecution, which was conducted by the (supposedly independent) Director of Prosecutions, lawyers representing various Commonwealth agencies (such as ASIO) repeatedly blocked efforts by the accused to have the functions of Pine Gap discussed in open court. When the accused tried to refer to the already published findings of the Joint Parliamentary Committee on Treaties, Commonwealth lawyers invoked a piece of legislation from 1987 preventing anything said in parliament from being repeated in a court.

The election of a Labor government committed to pulling Australian troops out of Iraq undermined the Crown's claim that the public conscience had been 'shocked' by the lenient sentences meted out to peaceful trespassers challenging Australia's participation in the war. In the end it didn't matter because in February 2008 the Northern Territory Court of Criminal Appeal quashed all convictions under the 1952 Act, leaving the Pine Gap 4 guilty only of damage to property.

From a legal point of view, it was a significant victory for the protesters. The Crown's clumsy attempt to crush them with the arcane provisions of the *Defence (Special Undertakings) Act* turned into a debacle. But the purpose of the break-in had been to put Pine Gap itself on trial, and in this the protesters failed. Entry to Pine Gap remained prohibited and so did public discussion of its operations, even when based on information already tabled in parliament. Bolt cutters had got the Christian activists through the perimeter wire but the secrets of Pine Gap were as impenetrable as ever—or so it seemed.

In court, the Pine Gap 4 had challenged the Crown to prove that Pine Gap was necessary for the defence of Australia. The Howard government refused to be publicly interrogated on this subject either in a court of law or anywhere else. A year after the quartet was freed by the Court of Criminal Appeal, the government quietly slipped an amendment to the *Defence (Special Undertakings) Act* into a miscellaneous defence bill, formally defining Pine Gap as a 'special defence undertaking' and a 'prohibited area' necessary for the defence of Australia. The amendment pushed the joint facility further beyond the scrutiny of the Australian parliament. The defence minister, Joel Fitzgibbon, said it would 'deter mischief-makers and those with more sinister intent'.

But between the appeal court's ruling and the government's amending of the law, rents had begun to appear in the shroud of secrecy that had enveloped Pine Gap since the 1960s.

On 15 October 2008 the US National Reconnaissance Office formally declassified the fact of its 'presence' at the Joint Defence Facility Pine Gap. The fact alone did not qualify as news: a quarter of a century earlier Des Ball had identified the NRO as the 'administrator' of the US base. The surprise was that the NRO itself had chosen to reveal it. After decades of almost pathological aversion to public disclosure, the organisation decided that the people did, after all, have a right to know. An internal 'questions and answers' memorandum on the subject of 'mission ground station declassification' proposed the following answer to the hypothetical question 'What do you do at this site?':

> The NRO supports the joint missions at JDFPG and RAFMH [RAF Menwith Hill in Yorkshire] through the provision of technical systems and shared research and development. The NRO's participation is achieved with the consent of the Host governments and contributes to the national security of the countries involved.

Cautious acknowledgment of already known facts was one thing: even a cursory study of press reports and other publicly available documents would have uncovered the central role of the NRO in the US spy satellite program. But candid exposure of classified and unadmitted material was another. When it came to discussing what really went on at Pine Gap, the gag remained as tight as ever. In response to the hypothetical question 'Can I talk about what happens at these sites?', the answer to NRO staff was an emphatic 'No':

> You may not discuss the plans, status, operations, and details of our satellites or our work. Don't discuss classified information such as:

- Intelligence information/data
- Specific details of partnerships with foreign entities
- Operations
- History of NRO MGS [mission ground station] missions and facilities
- Specifics of MGS staffing
- Budgets, costs, or expenditures

In short, nothing had really changed. The NRO's new policy of openness applied only to secrets that had long ceased to be secret. If the Australian public was ever to learn the truth about Pine Gap's role in US military operations, it would not be through the actions of Christian peace activists or Washington bureaucrats.

Edward Snowden was born in 1983 and grew up in the suburbs of Maryland, not far from the headquarters of the hyper-secretive National Security Agency, which had been intrumental in setting up Pine Gap and still had control over its spy satellites. Snowden's maternal grandfather worked for the FBI and was in the Pentagon on the day of the 9/11 attacks. Other members of his family also had jobs with the federal government. Snowden himself was by his own admission a patriot, with faith in American righteousness.

According to an article in *Wired* by the NSA expert James Bamford, Snowden was on his way to the office when the 9/11 attacks took place. 'I was driving in to work and I heard the first plane hit on the radio,' Snowden told Bamford. Later, after the US had dragged an international coalition into the Iraq war, Snowden recalled being 'very open to the government's expla-nation—almost propaganda—when it came to things like Iraq, aluminum tubes, and vials of anthrax . . . I still very strongly believed that the government wouldn't lie to us, that our government had noble intent, and that the war in Iraq was going to be what they said it was, which was a limited, targeted effort to free the oppressed.'

Snowden's idealism did not last. After an ill-fated attempt to become a special forces soldier—he broke both his legs in a training accident—he was hired by the CIA. By then he was already becoming disillusioned about the purpose of the war he had previously supported.

Snowden saw himself as a computer wizard; the CIA appeared to agree and sent him to its secret high-tech training school. In 2007 he was posted to Geneva, where he worked as a telecommunications information systems officer, as well as looking after the heating and air-conditioning. While in Switzerland, according to Glenn Greenwald's book *No Place to Hide: Edward Snowden, the NSA and the surveillance state*, he was regarded as the 'top technical and cybersecurity expert' in the country and was 'hand-picked by the CIA to support [President Bush] at the 2008 NATO summit in Romania'. The more Snowden saw of how the CIA operated, the more disillusioned he became. By the end of 2009 he had made up his mind to quit and was beginning to 'contemplate becoming a whistleblower and leaking secrets that he believed revealed wrongdoing'.

After leaving the CIA, Snowden worked for two contractors to the NSA. Sent to an NSA facility at Yokota Air Base near Tokyo, Snowden taught senior officials and military officers how to keep their networks safe from Chinese hackers.

For a while he trusted the new president, Barack Obama, to put a stop to the abuses he had seen being carried out in the name of the so-called 'war on terror', such as human rights violations at Guantanamo Bay and the mass surveillance of private communications. But far from discontinuing the programs that Snowden objected to, the Obama administration appeared to be extending them. Snowden realised, he told Greenwald, that the NSA and other US intelligence agencies were working with Silicon Valley corporations to build a system 'whose goal was the elimination of all privacy'.

In his book *The Snowden Files: The inside story of the world's most wanted man*, Luke Harding writes that Snowden had 'lost faith in meaningful congressional oversight of the intelligence community'. Any doubts were dispelled by the sight of the director of national intelligence, James Clapper, brazenly assuring the Senate intelligence committee in March 2013 that the US government did 'not wittingly' collect data on millions of Americans. It was a lie, and Snowden knew it.

The same month, Snowden moved from Dell Corporation to another NSA contractor, Booz Allen Hamilton, where he was able to read classified files confirming the extent of the agency's global surveillance programs. As a system administrator, Snowden had virtually unlimited access to NSA data and he began to download thousands of classified documents onto thumb drives, which he sent to carefully vetted journalists, including the US freelancer Glenn Greenwald. Given a month's leave to obtain treatment for his epilepsy, Snowden fled the United States in May 2013. He was holed up in a Hong Kong hotel when articles based on his leaked NSA data began to be published by newspapers around the world.

In the following months media reports in the US, the UK, Germany, France, Brazil, Australia and other countries revealed the scale and penetration of the global surveillance apparatus operated by the United States and its partners in the Five Eyes network, including Australia. According to Glenn Greenwald, Snowden handed over between 9000 and 10,000 documents, although US officials accused him of having stolen as many as 1.7 million documents. In November 2013 the editor-in-chief of Britain's *Guardian* newspaper claimed that only 1 per cent of the leaked NSA data had so far been published.

Australian officials claimed that Snowden had leaked up to 20,000 Australian files. Among the embarrassing revelations was the fact that the Australian Signals Directorate had tried to bug the phone of Indonesia's President Susilo Bambang Yudhoyono and had monitored his calls for fifteen days in 2009.

According to Greenwald, the 'vast majority' of the NSA files stolen by Snowden were classified 'top-secret'. While some were marked 'NOFORN' (no foreign distribution), most were designated 'FVEY', meaning they were approved for circulation to other members of 'Five Eyes', the network through which the NSA conducted the bulk of its surveillance work—much of it channelled through Pine Gap.

Once a hub of the CIA's global signals-gathering apparatus, Pine Gap had since the 1990s been fully integrated into the Pentagon's war-fighting machinery. In their 2015 paper 'The militarisation of Pine Gap: Organisations and personnel', Des Ball, Bill Robinson and Richard Tanter noted that units of all four branches of the US armed forces were based at Pine Gap and that US military personnel 'now comprise about 66 per cent of the US Government employees' at the base.

During the 1990s US military personnel had helped with processing and analysing signals intelligence intercepted by geosynchronous satellites. Since then, they had taken on a different role, euphemistically described as 'information operations, cyber warfare' and the achievement of 'information dominance'.

In practice, as Ball, Robinson and Tanter explained, this meant 'monitoring Internet activities being transmitted via communications satellites, scouring e-mails, Web-sites and Chat Rooms for intelligence to support military operations', particularly those involving special forces in Iraq and Afghanistan.

David Rosenberg, who witnessed the arrival of the first US military contingent at Pine Gap in late 1990, notes in his book that US and Australian military personnel provided useful support for the civilians who had previously performed all operational tasks. They brought a 'different perspective' to the civilian staff, Rosenberg says, as they were 'trained in warfare'.

By analysing the promotions of military officers employed at Pine Gap, Ball, Robinson and Tanter demonstrated how a spell in Alice Springs could be a springboard for careers in all four arms of the US military. They also found 'pathways that link Pine Gap to postings in combat zones in Iraq, Afghanistan, and the rapidly developing Africa Command, as well as to other key SIGINT facilities and cryptological or cyber units in the US . . . and in Japan and South Korea'. Not all these pathways led towards the Pentagon's 'official' wars in the Middle East; at least one led to 'drone operations . . . outside legally sanctioned war zones'.

There was a reason, of course, why Pine Gap staff might be recruited to work in drone operations: they already had experience. Pine Gap supplied intelligence that was used by the US military to identify and locate targets for assassination by armed drones or special forces. Alleged terrorists had been killed not only in Afghanistan but in countries such as Pakistan, Syria, Yemen and Somalia, where the US was not officially at war. Such attacks had been blamed for hundreds of civilian deaths.

Pine Gap's role in the Pentagon's lethal drone strikes had never been admitted by either government, although it could be deduced from articles written by Des Ball and others, from newspaper reports and from David Rosenberg's book. The proof was buried inside the cache of files stolen by Edward Snowden, but how long would it take to find it?

Chapter 26
Rainfall

In August 2017, ABC Radio's *Background Briefing* published five leaked NSA documents on its website, all classified either 'secret' or 'top-secret'. All had been submitted to the NSA for comment before publication. Most were redacted.

An NSA information paper, dated April 2013 and classified 'top-secret', had confirmed that the Joint Defence Facility Pine Gap—referred to by its unclassified NSA codename, Rainfall—played a 'significant role in supporting both intelligence activities and military operations'.

An NSA 'site profile', also classified 'top-secret', described Pine Gap as a 'unique facility' jointly staffed by both US and Australian personnel. The US contingent, it said, 'comprises NSA/CSS civilian, contractor, and Service Cryptologic Component (SCC) personnel from the US Army, Navy, and Air Force, as well as civilian, military, and contractor personnel from the CIA and NRO'. Significantly, it noted that Pine Gap's role went far beyond the 'collection' of signals intelligence. Rainfall, it said, 'detects, collects, records, processes, analyzes, and reports on PROFORMA signals collected from tasked target entities'. One of its 'primary mission areas' was the 'detection and geolocation of COMINT [communications intelligence], ELINT [electronic intelligence] and FISINT [foreign instrumentation signals intelligence]' and the site 'has a number of tools available for performing geolocations'. Under 'major RAINFALL mission components', the document listed 'Support to Military Operations (SMO)'.

Behind the acronyms and military euphemisms lay an inescapable fact: intelligence collected and analysed at Pine Gap was used in US combat missions, including drone strikes. If the US had the blood of innocent people on its hands, so did Australia.

The oldest of the documents published by *Background Briefing*, a 2005 secret 'program overview' of a new mission designated 'Mission 8300', confirmed much of what Des Ball had been saying and writing for more than a decade.

The Mission 8300 system consisted of 'four satellites in near-geosynchronous earth orbits' offering 'stable, continuous dwell for 24-hour collection'. Primary target areas were the 'former Soviet Union, China, South Asia, East Asia, Middle East, Eastern Europe, and the Atlantic landmasses'. Among 'system missions', the document listed the following:

- SIGINT support to US military combat operations;
- crisis monitoring;
- indications and warning support to the United States and deployed US forces;
- COMINT associated with command and control of military forces, movement of VIPs, deployment of military units, states of readiness, training proficiency, and combat operations;

- monitor testing activity to detect changes in weapons employment doctrine and to verify compliance with strategic arms limitation agreements; and
- monitor nuclear weapons and high-energy weapons testing.

The purpose and operating secrets of Rainfall were concealed behind an official 'cover story' which stated that Pine Gap was a 'joint US/Australian defence facility whose function is to support the national security of both the US and Australia. The JDFPG contributes to verifying arms control and disarmament agreements and monitoring of military developments. The JDFPG is jointly staffed by US and Australian DoD civilians and members of the various military branches.'

The leaked NSA documents not only blew the cover story, they also exposed the Orwellian syntax of evasion and misinformation that supported it. While the cover story was unclassified, the fact that the NRO 'uses cover stories' was secret. The term 'Joint Defence Facility Pine Gap' was unclassified, but the term 'Australian Mission Ground Station' was secret. The fact that the NRO had a ground station at Pine Gap was secret (it would not be admitted until 2008) and so was the association of CIA personnel with the base. The presence of the NSA, like the codename Rainfall, was unclassified. The fact that NSA activity at Pine Gap supported 'specific military operations' such as Operation Iraqi Freedom was unclassified provided 'no details of that support are revealed'. Support for NATO operations, however, was secret.

As well as confirming Pine Gap's involvement in US military operations, the stolen documents revealed that Australia was a valuable spying partner in the Pacific, Southeast Asia and China. According to the April 2013 information paper, 'close collaboration' between the NSA and Australia's Defence Signals Directorate 'has been particularly useful in providing cryptologic insight into Chinese targets'. Australian spying on China, the paper said, was 'already significant and will increase... as it draws down its presence in Afghanistan'.

Under the heading 'Problems/Challenges with the Partner', the paper reported, 'None'. The years of antagonism with Gough Whitlam and the Labor left over Pine Gap were old history. The United States could trust its once fractious junior partner completely.

In Australia's national parliament, Pine Gap has become a non-issue, rarely mentioned except in a spirit of bipartisan agreement. On 20 February 2019 the defence minister, Christopher Pyne, told the House of Representatives that the joint facility had 'evolved from its original Cold War mission... to meet new demands and new challenges' for which it had 'acquired cutting-edge, innovative technologies'. Repeating the line used by his predecessors about Pine Gap's role in 'arms control' and 'counterproliferation', Pyne assured the parliament that the base made a 'crucial contribution to global stability'. On its more shadowy role of 'delivering intelligence on a range of contemporary security priorities', the defence minister was more circumspect, noting that 'as a matter of longstanding practice' the government did not comment on intelligence matters while insisting that 'Australians can be assured that the government has full oversight of activities undertaken at Pine Gap'. Responding for the opposition, Richard Marles could only say that 'Labor supports every word the minister has said'.

While the era of mass demonstrations outside the gates of Pine Gap is long gone, anti-war protesters occasionally manage to infiltrate the heavily guarded base. In November 2017 six

Christian 'peace pilgrims' were found guilty of entering a prohibited area after breaking through the perimeter wire. One of the group, Margaret Pestorius, played her viola as she was arrested.

Snowden's disclosures, and those of other whistleblowers such as Chelsea Manning, revealed dark truths about US militarism in the 21st century: the brutalisation and torture of prisoners; extra-judicial killings by remote control; cover-ups of civilian deaths. The White House's 'war on terror' was exposed as a euphemism for dirty wars, declared and undeclared, waged around the globe with high-tech weapons, in which Pine Gap played a vital role.

Des Ball knew more about Pine Gap than anyone who had not worked there, and perhaps more than many who had. For most of his life he supported its presence on Australian soil, believing that, on balance, the benefits outweighed the risks. Yet towards the end of his life Ball admitted to having changed his mind, telling the ABC in 2014 that he had 'reached the point now where I can no longer stand up and provide the verbal, conceptual justification for the facility that I was able to do in the past'. In the world of modern weaponry it was no longer possible, Ball said, to distinguish between intelligence and operations. As a 'key node' in America's military surveillance network, Pine Gap was part of the Pentagon's war machine—a war machine doing things that Ball found 'very, very difficult . . . as an Australian, to justify'.

Ball was always sceptical of the 'jointness' of the joint facility. In his final interview, published in the *Saturday Paper* a few weeks before he died in October 2016, Ball told Hamish McDonald that Australia got '[e]verything, and nothing' from its joint investment in Pine Gap. 'Everything, in the sense that we get access to all this intelligence flowing through. Nothing, in the sense that it's not really what we want.' The carefully worded conclusion he reached in *Pine Gap: Australia and the US geostationary signals intelligence program* remains largely true today:

> Although there is now full Australian participation in the operations at Pine Gap, it remains the case that the station is essentially a US facility; it would not have been established except to satisfy US strategic interests; and very little of the intelligence collected at the station is of direct relevance.

In the words of Ball's colleague Richard Tanter, Pine Gap 'is, as it always has been, an American base, to which Australia is granted some access'. In his paper 'The "Joint Facilities" revisited', Tanter writes that 'the great expansion of the facility … has been driven by American concerns, built by the United States, paid for by the United States, and commanded by the United States. Were the United States to withdraw from its activities at Pine Gap, there would be nothing autonomous of significance left for Australia to operate.'

Chapter 27
A saucerful of secrets

Pine Gap had its origins in the strategic and scientific preoccupations of the US government in the two decades after the Second World War, in particular the nuclear arms race and the parallel race to militarise space. The CIA's Directorate of Science and Technology was at the forefront of research into space-based surveillance systems designed to give the US early warning of Soviet missiles. But in the minds of many Americans, there were more sinister threats than the Soviets. The CIA also monitored reports of unidentified flying objects. These reports came not just from the United States but from all over the world. Sightings were especially prevalent in eastern Europe and Afghanistan, prompting fears within the US military that the Russians were experimenting with 'flying saucers'. But there was another possibility that could not be conclusively ruled out by the agency's top scientific brains: that some UFOs were not man-made and might be the handiwork of an alien intelligence. In CIA-language, unidentified flying objects constituted a 'problem'.

The beginning of the Cold War coincided with (and probably inspired) a spate of UFO sightings. The earliest report of a 'flying saucer' over the United States came on 24 June 1947, when a private pilot noticed nine disc-shaped objects in the sky near Mt Rainier, Washington, flying at an estimated speed of more than 1600 kilometres per hour. More sightings poured in from all over the country. The witnesses were often reputable citizens, some with technical knowledge of flying. They included civilian and military pilots and air traffic controllers. As a result the US Air Force set up Project Saucer (later renamed Project Sign) to gather and analyse all reports of UFOs. The premise, according to Gerald Haines's article 'CIA's role in the study of UFOs, 1947–90: A die-hard issue', was that the sightings might be genuine and 'of national security concern'.

Project Sign quickly concluded that nearly all UFO sightings could be attributed to mass hysteria and hallucination, hoax, or the misinterpretation of known objects. But it recommended continued military intelligence control over the investigation of UFO sightings and, significantly, left open the possibility that the cause of some sightings could be extraterrestrial.

The air force continued to collect information about sightings, while publicly playing down the likelihood of extra-terrestrial activity. Rumours of official military involvement in the investigation of UFO sightings only fuelled public hysteria.

In 1952 the air force's director of intelligence ordered a new UFO project called Project Blue Book, described by Haines as the 'major Air Force effort to study the UFO phenomenon throughout the 1950s and 1960s'.

Although the air force was chiefly responsible for investigating UFO sightings, other federal agencies were also involved.

For a time the FBI had been following up reported sightings, but this stopped when the bureau's chief, J. Edgar Hoover, read a restricted letter from the army indicating that FBI agents had merely been enlisted to relieve the air force 'of the task of tracking down all the many instances which turned out to be ashcan covers, toilet seats and whatnot'.

July 1952 saw a dramatic escalation in UFO sightings across the United States. On two consecutive days, radar screens at Washington National Airport and Andrews Air Force Base showed unusual blips. A week later the blips were seen again. Fighter jets were sent up to investigate but the pilots saw nothing. The mysterious radar blips were widely reported in the press. When the White House asked the air force for information, it was told the blips could be due to 'temperature inversions'—an explanation that was later found to be true but which did little to calm the public's nerves.

Unlike the FBI, the CIA was intensely interested in UFOs and studied the air force's reports of sightings. From the start, however, the CIA's involvement was to be kept secret. Worried by the public's 'alarmist tendencies' to interpret CIA interest in UFOs as proof of 'unpublished facts' known to the US government, a senior official in the agency's Office of Scientific Intelligence (OSI) recommended on 1 August 1952 that 'no indication of CIA interest or concern reach the press or public'.

The OSI concluded that the air force's investigations of UFO sightings were not stringent enough to be able to determine the nature of the 'flying saucers'.

According to a CIA briefing dated 22 August 1952, the air force had ruled out the possibility that flying saucers were either US secret weapons, Soviet secret weapons, or extraterrestrial visitors. Instead, the air force believed they were:

1. Well known objects such as balloons (over 4000 are released daily in the U.S.), aircraft, meteors, clouds etc not recognised as such by the observer.
2. Phenomena of the atmosphere which are at present poorly understood, eg refractions and reflections caused by temperature inversions, ionization phenomena, ball lightning etc.

Of the 'thousands' of reported flying saucer sightings, the briefing noted that the air force had explained 78 per cent as belonging to the two categories above and 2 per cent as hoaxes. That left 20 per cent of UFO sightings unexplained 'primarily because of the vague descriptions given by observers'. (Project Blue Book investigated more than 7000 UFO reports between 1953 and 1965 and attributed 80 per cent to 'natural phenomena, hoaxes, birds or man-made objects'; 17 per cent did not provide sufficient data for thorough analysis and 3 per cent were classed as 'unidentified'. By the time Blue Book was disbanded in December 1969, 710 sightings out of a total of 12,618—just over 5 per cent—remained 'unidentified'.)

According to the CIA, the air force was 'mostly interested in the "saucer" problem because of its psychological warfare implications'. A study of publications designed for 'Soviet consumption' uncovered not a single reference to 'flying saucers'. In the case of internationally reported UFO sightings, the study 'found not one report or comment, even satirical, in the Russian press. This could result only from an official policy decision and of course raises the question of why and of whether these sightings could be used from a psychological warfare point of view either offensively or defensively.'

By contrast, publications intended for American readers were cultivating a flying saucer 'craze'. Key members of American 'saucer societies' had been 'exposed as being of doubtful loyalty... the societies, in some cases, are financed by an unknown source'. The implication was clear: the US Air Force suspected the 'flying saucer craze' was being covertly funded and fomented by the Soviets.

The study paid particular attention to one group, the Civilian Saucer Committee in California. It found the organisation 'has substantial funds, strongly influences the editorial policy of a number of newspapers and has leaders whose connections may be questionable'. The air force would continue to monitor the Civilian Saucer Committee 'because of its power to touch off mass hysteria and panic' while watching for 'any indication of Russian efforts to capitalise on this present American credulity'.

While its fears of 'mass hysteria and panic' were real enough, the air force was more concerned about the effect of the saucer craze on its air defence capabilities:

> The Air Force realizes that a public made jumpy by the 'flying saucers' scare would be a serious liability in the event of air attacks by an enemy. Air defences could not operate effectively if the air force were constantly called upon to intercept mirages which persons had mistaken for enemy aircraft.

Dissatisfied with the air force's efforts, the CIA was eager to build up its own expertise on the subject of UFOs. At a meeting of CIA branch chiefs on 11 August 1952, an official declared that 'a project is to be started in the P & E [processing and exploitation] division on "flying saucers"'.

The following month the CIA's deputy director of scientific intelligence, H. Marshall Chadwell, wrote a memo on the subject of 'flying saucers' to his boss, the director of the CIA, General Walter B. Smith. The 'problem', Chadwell wrote, was to determine:

a. Whether there are national security implications in the problem of 'unidentified flying objects' i.e. flying saucers;
b. Whether adequate study and research is currently being directed to this problem in its relation to such national security implications; and
c. What further investigation and research should be initiated, by whom, and under what aegis.

The seriousness of the flying saucer problem, Chadwell told Smith, 'transcends the level of individual departmental responsibilities, and is of such importance as to merit cognizance and action by the National Security Council'. He recommended that Smith enlist 'appropriate agencies' inside or outside the government to undertake 'the investigation and research necessary to solve the problem of instant positive identification of "unidentified flying objects"'. Second, he recommended the agency 'investigate possible offensive and defensive utilization of the phenomena for psychological warfare purposes both for and against the United States'. Third, he wanted the agency to provide the National Security Council with a 'policy of public information' that would 'minimize the risk of panic'.

Senior figures in the CIA were now convinced that the 'flying saucer problem' represented a grave threat to national security. 'At any moment of attack,' Chadwell told Smith, 'we are now in a position where we cannot, on an instant basis, distinguish hardware from phantom, and as tension mounts we will run the increasing risk of false alerts and the even greater danger of falsely identifying the real as phantom.'

While the CIA's main concern was UFO sightings over the United States, declassified files show that the agency collected data about UFO sightings from all over the world. Under the subject heading 'military—unconventional aircraft', a document dated 24 November 1952 reported sightings from Spain, French Morocco, Tangier and French West Africa.

On 17 July 'disks' had appeared over Marrakesh:

> Marrakech (special correspondent)—At 2100 hours... many people saw a large, luminous disk flying horizontally, with a leaping and bounding motion. Then, there was a burst of light. A second disk of smaller dimensions appeared, going off horizontally toward the southwest. The whole appearance lasted about one minute.
>
> From several points of the Marrakech region, luminous disks were seen travelling at a dizzy speed.
>
> On 14 July, flying saucers had been seen over the Ilfrane region, flying toward Meknes.

Two months later local newspapers carried reports of a 'strange object' over Tangier and Fez:

> On 21 September 1952, at 1815 hours, many people at a beach 17 kilometres from Tangier saw toward the south a luminous disk of a diameter close to that of the setting sun, flying horizontally from east to west. After holding its course for about 12 seconds, the disk, which looked like a brightly illuminated metal object, suddenly emitted two great streaks of flame and disappeared.

At 1820 hours, a strange object was also seen in the sky over Fes, French Morocco, flying at high speed from east to west and leaving behind a luminous white trail.

By the end of 1952, the agency's office of scientific intelligence had begun setting up a panel of experts 'of sufficient competence and stature to ... convince the responsible authorities in the community that immediate research and development on this subject must be undertaken'.

Chadwell himself was reluctant to rule out the possibility of 'extra-terrestrial visitors', advising the CIA director in a memo in December 1952 that 'sightings of unexplained objects at great altitudes and travelling at high speeds in the vicinity of major US defence installations are of such nature that they are not attributable to natural phenomena or known types of aerial vehicles'.

The same month, the director of air force intelligence, Major General Samford, promised full air force cooperation with the CIA's proposed panel of distinguished non-military scientists to investigate the flying saucer problem.

Chaired by H.P. Robertson, a physicist and weapons expert from the California Institute of Technology, the panel convened in January 1953 and included Luis Alvarez, a physicist who would be awarded the Nobel Prize in 1968, as well as experts in nuclear physics, geophysics, radar and electronics. Its brief was to review the available evidence on UFOs and to determine whether UFOs represented a threat to US national security.

Like Projects Sign and Blue Book, the Robertson Panel found that most, if not all, sightings of 'flying saucers' had a rational explanation. Several sightings—notably at Tremonton, Utah, on 2 July 1952 and near Great Falls, Montana, on 15 August 1950—were analysed in detail. In its classified final report the panel unequivocally ruled out visits by extraterrestrials while warning that mass UFO reporting could swamp government communications and induce public hysteria, and that US air defences could be compromised. To minimise these risks it called for a public education campaign to debunk sightings of flying saucers:

> The Panel concluded unanimously that there was no evidence of a direct threat to national security in the objects sighted. Instances of 'Foo Fighters' were cited. These were unexplained phenomena sighted by aircraft pilots during World War II in both European and Far East theaters of operation wherein 'balls of light' would fly near or with the aircraft and maneuver rapidly. They were believed to be electrostatic (similar to St Elmo's fire) or electromagnetic phenomena or possibly light reflections from ice crystals in the air, but their exact cause or nature was never defined ... If the term 'flying saucers' had been popular in 1943–45, these objects would have been so labelled. It was interesting that in at least two cases reviewed the object was categorized by Robertson and Alvarez as probably 'Foo Fighters', to date unexplained but not dangerous; they were not happy thus to dismiss the sightings by calling them hoaxes.

As far as the CIA was concerned, the findings of the Robertson Panel laid the 'flying saucer' problem to rest. While the agency would continue to record sightings for national security purposes, it went cool on the idea of further investigation. Behind the scenes, the CIA went to great lengths to distance itself from the study of UFOs. The fact that the agency itself was behind the establishment of the Robertson Panel was officially covered up, along with evidence of its previous interest in flying saucers. Budgetary pressures were invoked as a reason for cutting back even the minimal resources that went into monitoring sightings.

But not everyone was prepared to let the matter drop. Chadwell, in particular, continued to be worried by UFO sightings, especially when reports emerged in the mid-1950s of Soviet efforts to develop a flying saucer–like military aircraft. While inconclusive, some sightings were considered plausible enough to merit the attention of the agency's most senior officials. One sighting, by US Senator Richard Russell on a visit to the USSR in October 1955, aroused particular concern. After analysing accounts by Russell and his fellow travellers, the agency's assistant director of scientific intelligence, Herbert Scoville, decided that the UFO was probably a conventional jet aircraft in a steep climb. In a classified memo to the CIA director, however, his conclusions appear more equivocal:

> Further discussions will probably be required before the matter can be completely resolved. In the meantime, however, the evidence does not appear sufficiently firm to warrant the conclusion that the Soviets have developed and have in operation a radically new type of aircraft.

In fact, it was the CIA's success with its own experimental aircraft—the U-2 high-altitude spy plane—that led to the next spike in flying saucer sightings.

In the mid-1950s few commercial airlines flew at altitudes higher than 20,000 feet. The U-2 was designed to fly above 60,000 feet—high enough to stay beyond the reach of Soviet MiG fighters. Early models were painted silver (later ones were painted black) and reflected the rays of the sun, especially at dawn and dusk. U-2s flying over the United States were often mistaken for flying saucers, especially by the pilots of airliners flying from east to west. A CIA publication entitled *The CIA and the U-2 Program: 1954–1974*, published in 1998, describes what happened:

> When the sun dropped below the horizon of an airliner flying at 20,000 feet, the plane was in darkness. But, if a U-2 was airborne in the vicinity of the airliner at the same time, its horizon from an altitude of 60,000 feet was considerably more distant... being so high in the sky, its silver wings would catch and reflect the rays of the sun and appear to the airliner pilot, 40,000 feet below, to be fiery objects. Even during daylight hours, the silver bodies of the high-flying U-2s could catch the sun and cause reflections or glints that could be seen at lower altitudes and even on the ground. At this time, no one believed manned flight was possible over 60,000 feet, so no one expected to see an object so high in the sky.

According to Haines's article, agency officials who worked on the U-2 and its successor, the SR-71 Blackbird (which flew much higher and much faster), estimated that 'over half of all UFO reports from the late 1950s through the 1960s were accounted for by manned reconnaissance flights (namely the U-2) over the United States'.

In order to protect the ultra-secret U-2 project, false and misleading explanations were often put out by the air force to account for these erroneous flying saucer sightings—explanations that would fuel theories about a UFO cover-up by the US military and the CIA.

Suspicions of a cover-up were given extra impetus by the first head of Project Blue Book, Edward J. Ruppelt, the man credited with inventing the term 'unidentified flying object'. In his book *The Report on Unidentified Flying Objects,* Ruppelt wrote about the investigation of UFOs by the US Air Force between 1947 and 1955. His disclosure of the existence of the Robertson Panel prompted others to push for the release of all the government's files on UFOs. With the air force's role in flying saucer investigations now public knowledge, pressure mounted on the CIA to admit its own role, and to declassify and release the findings of the Robertson Panel. The CIA refused, agreeing only to release a 'sanitized' version of the report that omitted any mention of the agency or its part in setting up the panel.

In 1966, the same year that three men from the CIA's Directorate of Science and Technology, Bud Wheelon, Carl Duckett and Leslie Dirks, drove into the Australian outback to toast the beginnings of Pine Gap, Republican congressman (and future president) Gerald Ford called for a congressional investigation into UFOs. Although quick to point out that he had 'never said that I believe any of the reported UFO sightings indicate visits to earth from another planet', Ford insisted that the American public 'deserves a better explanation [of the UFO phenomena] than that thus far given by the Air Force'.

> Those who scoff at the idea of a congressional investigation of UFOs apparently are unaware that the House Armed Services Committee has scheduled a closed-door hearing on the matter Tuesday with the Air Force and that Rep. Joseph E. Karth, D. Minn., headed a three-man subcommittee which held two days of hush-hush hearings five years ago on

behalf of the House Science and Astronautics Committee. Karth has confirmed in conversation with a member of my staff that he conducted these secret hearings.

The 'closed-door' hearing of the House Armed Services Committee that Ford referred to was uneventful, with the secretary of the air force, Harold Brown, maintaining that there was no evidence of visits to Earth by 'strangers from outer space'. Nevertheless, Brown insisted that the air force would keep an 'open mind' on the subject of UFOs and would continue to investigate all sightings.

Ford did not get the congressional inquiry he wanted, but his demands for the air force to come clean on what it knew about UFOs resulted in an air force grant of $300,000 for scientists at the University of Colorado to carry out 'independent investigations' into UFO sightings. The project was led by a physicist, Dr Edward Condon, who claimed to be 'agnostic' about flying saucers and considered visits by extraterrestrials 'improbable but not impossible'.

Meanwhile Bud Wheelon, a veteran of the CIA's Office of Scientific Intelligence, continued to juggle his time between the agency's fledgling geostationary satellite program and its investigation of flying saucers. Declassified CIA files show Wheelon's signature on a 1965 memorandum addressed to the CIA director entitled 'Evaluation of UFOs':

The Office of Scientific Intelligence/DD S&T monitors reports of UFOs, including the official Air Force investigation reports, and concurs with the Air Force conclusions, which are unclassified and available to scientific investigators.

The accompanying table of monthly statistics for the year 1964 highlighted two categories of UFO report: 'astronomical' and 'other' cases. Under 'astro cases', it included 'meteors' and 'stars and planets'. Under 'other cases' it listed:

- Hoaxes, Hallucinations, Unreliable Reports and Psychological Causes;
- Missiles and Rockets;
- Reflections;
- Flares and Fireworks;
- Mirages and Inversions;
- Search and Groundlights;
- Clouds and Contrails;
- Chaff;
- Birds;
- Physical specimens;
- Radar analysis;
- Photo analysis;
- Satellite decay;
- Other.

According to the memorandum signed by Wheelon, 'Of some 532 UFO reports investigated during 1964, only 5% are classed as unidentified; 11 investigations are still pending'.

Although evidence of ongoing CIA interest in UFOs is clear from declassified files, the agency continued to deny its involvement. Asked again in July 1966 to declassify the Robertson Panel's now thirteen-year-old report, the deputy director of OSI, Karl Weber, reiterated the CIA's refusal to publicly admit its sponsorship of the panel.

By the end of the 1960s official interest in flying saucers had waned. Painstaking scientific analysis of UFO reports over the past two decades had produced no tangible evidence of anything sinister. With CIA help, the Condon committee carried out detailed technical analysis of UFO photographs. The results convinced the committee that further study of sightings was futile. Condon recommended the air force disband Project Blue Book. The group's findings were reviewed by experts from the National Academy of Sciences, who declared, 'On the basis of present knowledge, the least likely explanation of UFOs is the hypothesis of extra-terrestrial visitations by intelligent beings.' Once again, the CIA's role was covered up.

During the 1970s conspiracy theories about the CIA's hoarding of UFO information multiplied along with the white radomes outside Alice Springs. The agency's involvement with flying saucers, like its involvement with Pine Gap, was scrupulously concealed. But the truth about both was about to be revealed.

In July 1975, a month after the launch of the new Argus spy satellite controlled from Pine Gap, and three months before Gough Whitlam found out that the facility was run by the CIA, not the Pentagon, the head of an Arizona-based flying saucer society accused the agency of covering up its involvement with UFOs. The agency falsely responded that 'at no time prior to the formation of the Robertson Panel and subsequent to the issuance of the panel's report has CIA engaged in the study of the UFO phenomena'. It went on to imply (again, falsely) that the CIA possessed no more UFO information. Three years later, demands made under American freedom of information laws revealed the existence of the agency's flying saucer files. According to Haines's article, a thorough and at times bad-tempered search 'finally produced 355 documents totalling approximately 900 pages'. All but 57 documents were released.

The mass release of previously hidden and denied documents did not satisfy everyone. The agency was accused of having withheld significant documents, of having made arbitrary deletions, and of having failed to conduct a proper search of its archives.

The *New York Times* and others repeated allegations of a major cover-up while significantly overstating the CIA's UFO activities. Victor Marchetti, the ex-CIA official who had helped draft the secret Pine Gap treaty between the US and Australian governments, described the opening of the CIA's UFO files as having 'the same aroma of the agency's previous messy efforts to hide its involvement in drugs and mind-control operations, both prime examples of a successful intelligence cover-up'. The new CIA director, Stansfield Turner, was said to have been taken aback by the criticism. In truth, the agency's interest in UFOs had for more than two decades been low-key and reactive, incited by waves of public sightings, rather than organised and strategic. The long-denied files, however, were real.

The cloak of secrecy the CIA had thrown over its flying saucer investigations also covered its spying operations at Pine Gap, and with similar results: in both cases, the public suspected the agency was up to more than it would admit. The mystery surrounding both UFOs and Pine Gap convinced many Australians that the two were linked.

Newspaper reports of the 1966 hearings of the US Congress's House Armed Services Committee, and of Condon's UFO study at the University of Colorado, aroused much interest in Australia, where the air force had responsibility for investigating UFOs.

The National Archives in Canberra contain hundreds of

reports of UFO sightings in Australia and its former territories, notably Papua New Guinea. In an attempt to counter the mysterious and sensational implications of the term 'unidentified flying objects', the Department of Air labelled them 'unidentified [or 'unusual'] aerial sightings'.

According to an air department memorandum about 'unusual aerial sightings', the RAAF spent 'a considerable amount of effort . . . investigating each report and the majority of observers are interviewed by selected RAAF personnel'.

Between 23 January 1960 and 30 December 1971, the RAAF received 595 UFO reports. Of these, the department assessed that '93 per cent were explainable by present scientific knowledge. Six per cent of the reports did not provide sufficient information to permit proper analysis and evaluation. One per cent of reports were attributed to unknown causes.'

In a 1965 speech to the Ballarat Astronomical Society, a member of the air department's operational research office, Mr B.G. Roberts, claimed that RAAF investigations of UFO sightings in Australia produced broadly similar results to those in the United States and the United Kingdom: around 90 per cent could be explained. As for the rest:

> I would like to make it clear that the Department of Air never has denied the possibility that some form of life may exist on other planets in the universe. Just as we on earth are at the brink of our entry into space, it is not impossible that somewhere else in the universe (<u>if intelligent life does exist out there</u>), others have or are about to do the same. However, the department has, so far, neither received nor discovered in AUSTRALIA any evidence to support the belief that the earth is being observed, visited or threatened

by machines from other planets. Furthermore, there are no documents, files or dossiers held by the department which prove the existence of 'flying saucers'.

It was noticed that meteor showers produced an upsurge in UFO sightings. Inundated with reports of flying saucers (some with 'humanoid' passengers) from all over the country, the Department of Air in Canberra sought help from the Commonwealth Scientific and Industrial Research Organisation (CSIRO) to evaluate sightings. The results were collated by the air department in typewritten summaries such as these dating the sightings and giving a brief description and a possible cause.

12 Feb 69	Time 2015 hrs. Duration 5–10 minutes, weather cloudy. Two objects, one bright green and one flashing. Melbourne, Vic.	Aircraft
15 Feb 69	Time 2045 hrs. Duration 5–10 minutes, weather fine. Big bright star, moving NW to SE at Bondi, NSW.	Echo II satellite
17 Feb 69	Time 0250 hrs. Duration 25 minutes, weather clear moonless. Observer attracted by noise then saw solid object and humanoid form walking around and heard morse code, at Flinders Park, SA.	Investigation has found no supporting evidence
18 Feb 69	Time 2005 hrs. Duration 25 minutes, weather clear. An object with changing red to white light occasionally stopped, moved S to N. Viewed from moving car at Melbourne, Vic.	Aircraft

21 Feb 69	Time 2125/2130 hrs. Duration 1½ seconds, weather slight overcast. Light moving quickly at Melbourne, Vic.	Meteorite
23 Feb 69	Time 2030 hrs. Duration 30 minutes, weather becoming cloudy. Three objects, one brilliant white. At Melbourne, Vic.	Combination of Echo II and aircraft
23 Feb 69	Circle of 'scorched' grass, burnt and squashed toadstools, at Kyogle, NSW.	Toadstool 'ring phenomena'
27 Feb 69	Time 0350 hrs. Duration 10 minutes, misty rain. Noise similar to aircraft, moving W at Aitape, TPNG [Territory of Papua New Guinea].	Aircraft
5 Mar 69	Time about 0915 hrs. Duration 15 seconds, cloudy. Burning object, red/orange, moving N to W with trail of smoke at Maryville, Vic.	Meteor
7 Mar 69	Time 2046 hrs. Duration 2 minutes, cloudy but clear. Brilliant yellow light, appeared like aircraft landing light, at Darwin, NT.	Moon 'halo' phenomenon, reflected on low cloud
11 Mar 69	Time 2100 hrs. Duration 2–3 minutes, cloudy but clear. Yellow lights in formation, quite high at Darwin, NT.	Moon 'halo' effect

Far from treating UFO sightings as the delusions of cranks and time-wasters, the authorities took them seriously, with information being passed up the chain, even landing on the desks of government ministers. After a Tasmanian clergyman, Reverend Lionel Brown, reported seeing eight flying saucers and a grey, cigar-shaped 'mother ship' over Cressy in

October 1960, the Minister for Air, Frederick Osborne, told the parliament that an RAAF wing commander would be sent to investigate. (He added, however, that he was 'not prepared to make official documents available'.)

On 23 February 1969 the Brisbane *Truth* reported two flying saucer landings near Kyogle, New South Wales. A local woman, Mrs Gibbs, visited the site to confirm the landing for herself and found a ring of scorched grass and some burnt toadstools, one of which she enclosed with her letter to the CSIRO in East Melbourne. In the event of any 'public reaction', Mrs Gibbs promised to keep her discovery confidential 'if you wish it so'. The burnt toadstool was sent to the assistant government botanist, J.H. Willis, for analysis. On 25 March Mr Willis sent the results of his investigation:

> I have identified this toadstool as Lepiota gracilenta (Slender Parasol Mushroom) which occurs widely in moist grassy places almost throughout temperate Australia. The flattened 'gills' underneath certainly look scorched, but that could easily have been caused by some animal trampling the fungus and then leaving its cup upside down to bake out in the sun. Toadstools of this kind are known sometimes to form 'fairy rings'—perennial circles of fruiting bodies on lawns and open grasslands that widen slightly, year by year, as food reserves in the soil are exhausted inside the growing circle. Usually there is a distinct ring of dry dead grass immediately behind the circle of the fungi, and this might well serve to explain Mrs Gibbs's 'definite circle of scorched grass'. It is my belief that she was witnessing a very ancient and intriguing phenomenon among certain grass-inhabiting toadstools.

The Weapons Research Establishment in Salisbury, South Australia was another government organisation drafted in to help with UFO investigations. Between the late 1950s and mid-1960s it was heavily involved in the top-secret development and testing of Britain's Blue Steel nuclear guided missile. The expertise gained by the WRE in rocketry and advanced engineering made it an obvious choice to help with the technical analysis of UFO sightings.

On 8 April 1963, around 100 miles north of Broken Hill in outback New South Wales, a man named McClure discovered a shiny metal sphere roughly 14 inches (35 centimetres) in diameter, weighing 12 pounds (5.4 kilograms). According to a file in the National Archives, the sphere had been spotted from overhead by a plane crossing part of the outback that had been 'shunned by human beings for at least 50 years'. McClure turned his find over to the government for further investigation.

Nearly two months later, at a point about 60 miles (97 kilometres) from the site of McClure's discovery, another shiny metal sphere was found. Like its predecessor, it was perfectly smooth, brightly polished, and 'without any aperture of any kind to give access to the interior'. It was slightly larger than the original sphere, weighing more than eighteen pounds.

A fortnight later, near Muloorina, New South Wales, a third sphere was found. This one was smaller, about six inches in diameter and weighing seven pounds, with 'one small aperture, about half an inch in diameter, which enabled investigators to ascertain that the sphere was lined with lead'.

On 30 April 1963 the minister for supply, Allen Fairhall, told the House of Representatives that 'without any doubt . . . it came from a space vehicle of some kind. The plumbing to and from the ball had been fused, and as the contents may be of scientific interest the sphere has not yet been opened.' According to Fairhall, the peculiar find represented 'a million-to-one chance; that is, a piece of orbiting hardware has been able to survive the temperature met at the time of re-entry into the earth's atmosphere and reach the earth in a solid piece'. While Fairhall appeared confident that the spheres were man-made, Australian scientists were said to be 'puzzled' by the metal spheres and to have 'no idea what they were or where they came from'. Efforts to open one of the spheres with drills and hacksaws were described as 'futile'.

A Department of Air minute paper written in August 1971 reveals that the Weapons Research Establishment was never able to solve the mystery of the metallic spheres. A 'safety officer' at the WRE remembered testing the spheres eight years earlier. According to the minute, the 'origin of these metallic balls was never finally determined. However, it is thought probable that they were part of the pressuring system of a satellite or satellite launcher, which made its re-entry over the Australian continent.' The same safety officer recalled testing another mysterious metallic object found at Shepparton, Victoria in 1953 or 1954 which turned out to be 'part of an early type of refrigerator'. If the aliens had landed, it appeared they had brought their whitegoods.

Three years after answering parliamentary questions about the mysterious metallic spheres, Prime Minister Holt made Fairhall his minister for defence. Fairhall was now in charge of Pine Gap. That meant he was in charge of keeping the secrets of Pine Gap, not just from his fellow members of parliament but from the Australian people.

As Holt's spokesman on unexplained objects dropped from space, Fairhall spoke candidly but failed to dispel suspicions that the government had something to hide. Flying saucers were not unknown over Canberra. On 28 April 1961 the *Canberra Times* had reported 'several' sightings of a 'mysterious silver object' passing over Canberra the previous afternoon. Callers described it as a 'glittering silver ball travelling at terrific speed'. A spokesman for the Department of Civil Aviation suggested it was probably a commercial airliner, distorted to the human eye by a 'strong setting sun'. In June 1962 an Ainslie man reported seeing a 'brilliant white light travelling across the sky north of the city'. According to the paper, 'Records existed of a number of sightings of "unidentified flying objects" for which investigations had failed to provide any satisfactory explanation.'

The following year the *Canberra Times* reported claims by an electrical fitter to have taken photographs of a 'huge saucer-like object' as it hovered over the southern part of Canberra before veering off towards Queanbeyan. A spokesman for the nearby Mt Stromlo Observatory said there had been no reports of unusual sightings, while the RAAF and the control tower at Sydney airport denied there were either military or civilian planes in the area. The paper published the photographs, together with the comment that 'No official explanation of the phenomenon has been made.'

In June 1966 the same newspaper reported that two police constables in a car had 'chased' a flying saucer which had been seen by 'dozens of people' over Grafton in New South Wales. The saucer 'moved slowly at about 1,500 ft above the ground and . . . changed colour every few minutes. The colour alternated between white and red.' The same saucer had been seen over Lismore, 130 kilometres away. Meteorological officials suggested that the saucer was in fact the star Canopus.

A year later, Australian newspaper reporters began asking questions about Pine Gap. The press as a whole proved surprisingly compliant with the defence minister's desire for secrecy, 'collaborating' with the government (as the left-wing *Tribune* newspaper put it) by 'averting their eyes'.

Some Australians who had followed newspaper reports of the UFO investigation at the University of Colorado and knew about Project Blue Book found parallels between the secrecy surrounding UFOs and the secrecy surrounding Pine Gap.

In the late 1960s and early 1970s rumours began to circulate among UFO enthusiasts about a string of unexplained UFO sightings. There were allegations of UFO 'crashes' or 'explosions' at Church Point in Sydney in April 1969; at Indooroopilly, Queensland in December 1970; and at Nowra, New South Wales in 1976. In October 1972 an Albury man named Norman Benstead claimed to have taken photographs of a UFO flying over Lake Hume. The sighting was not reported to the RAAF, which had no reports of sightings in the Albury area during September, October and November 1972. It was reported, however, to the Unidentified Flying Objects Investigation Centre, a private UFO society based in Sydney. The society sent Benstead's photographs to the CSIRO for analysis. Officially, the CSIRO turned down the request, but one CSIRO scientist was intrigued enough to examine the pictures. He worked out that they were fakes, created by winding back the film in a cheap instamatic camera to allow a double exposure. To prove Benstead's pictures were fakes, the scientist produced identical pictures himself, which he sent to the UFO investigation centre, along with the suggestion that the worthless negatives be destroyed.

Six years later, the story of Benstead's alleged UFO sighting surfaced again in media reports carried by the Melbourne *Herald* and Channel 7. According to the *Herald*, 'The CSIRO spent weeks examining the negative and double-exposed a film using the top of a lamp shade superimposed above a photograph of an office block to try to recreate the effect of the saucer-shaped object in Mr Benstead's photograph.' Benstead himself pleaded ignorance, telling the paper, 'When I got the report back saying it was a lamp shade, I couldn't be bothered following it through with publicity, though I'm blowed if I know how I double-exposed the negative.'

Not everyone accepted that the Benstead photographs were fakes. Allegations that the CSIRO scientist had recommended the negatives be destroyed fuelled suspicions among UFO enthusiasts of another cover-up.

In March 1981 the fate of the photographs formed part of a long letter to the Coalition defence minister, Jim Killen. This letter also referred to other UFO sightings, including one over Port Moresby in August 1953, colour film of which had reportedly been sent to the US Air Force for evaluation; a suspected UFO 'landing' at Clayton, Victoria in April 1966; and the disappearance of Frederick Valentich, a young Australian pilot and flying saucer enthusiast who disappeared while flying over Bass Strait on 21 October 1978. UFO watchers in both Australia and the United States claimed that a UFO had caused him to crash, although a more likely explanation was that the inexperienced pilot became disoriented by bright lights above him (probably the planets Venus, Mars and Mercury and the reddish star Antares) before crashing into the sea.

After asking Killen for details of Australian involvement in Project Magnet (a Canadian UFO study dating back to the early 1950s), the letter went on:

Re Pine Gap:

What was the function defined by the USA in seeking to establish the base. Was it a joint Australian-USA Defence space research facility?

If a space research facility only, why is it so top secret that access to Australian parliamentarians is denied?

Is there an underground city at Pine Gap?

Are UFO/USO [unidentified submerged object] experiments (antigravity-electrogravitic propulsion) conducted there or from there, and is Australia actively involved in such experiments?

A thick file at the National Archives entitled 'Miscellaneous enquiries—General—Unidentified Flying Objects—UFOs' shows that the letter reached the office of not just the defence minister but the minister for science and technology and the minister for administrative services. The subject was considered sensitive enough for the action sheet to be marked 'Please pass by hand at every stage'.

There is no record of how (or even whether) Killen answered the questions about Pine Gap. But in the minds of some Australians, Pine Gap was now firmly linked to covert US military research possibly involving UFOs.

Since then the links have become stronger. The internet is littered with websites detailing strange happenings inside and underneath the unmarked buildings at Pine Gap. True and almost true statements about the facility ('Control centre for spy satellites which circle the globe 24/7', 'Provides satellite tracking for the secret space program') sit alongside UFO mumbo-jumbo ('Meeting place for ETs & secret govt', 'Military/alien abduction receiving facility', 'Time portals—Earth energy and grid lines').

According to one website, 'Pine Gap has enormous computers which are connected to their American and Australian central counterparts, which collect all the information secured in these countries, not only about finance and technology, but on every aspect of the life of the average citizen [an assertion that Edward Snowden has shown to be largely accurate]. Those computers at Pine Gap are also evidently connected to similar mainframes in Guam, in Krugersdorp South Africa, and at the Amundsen-Scott US base at the South Pole.'

The same website claims that 'the most disquieting fact about Pine Gap may be that the employees working on the base, and especially those earmarked for duty on electromagnetic propulsion projects, have undergone brainwashing and even implantation of intracranial devices. Those employees have turned into unconditional slaves of their master, whoever he is.'

Pine Gap is also the setting for a 2012 movie called *Crawlspace*, directed by former make-up man and special effects supervisor Justin Dix. According to the synopsis:

> Deep in the heart of the unforgiving Australian desert lies Pine Gap, a top-secret government facility operated by the United States military. When the base comes under attack from unknown forces, an elite team is sent in to extract the military scientists.

Variously described on the review website Rotten Tomatoes as a 'B-movie horror/Sci-fi mash up', a 'sloppy mess' and a 'tight 80-minute ride of hallucinations and chilling medical discoveries', the movie referenced Pine Gap in its promotional material and incorporated long shots of the radomes, although it was actually shot at Melbourne's Docklands Studios. Dix's fictional

'Pine Gap' bore little resemblance to the real base but tapped into a mythology of sinister goings-on nourished by decades of government duplicity.

The road still bristles with signs warning unwanted visitors to keep away, but the most serious threats to the secrets of Pine Gap have rarely entered by the front gate.

Bibliography

ABC, 'The Base: Pine Gap's role in US warfighting', *Background Briefing*, 20 August 2017, https://www.abc.net.au/radionational/programs/backgroundbriefing/the-base-pine-gaps-role-in-us-warfighting/8813604 (accessed 9/05/2019)

Andronov, A., 'American Geosynchronous SIGINT Satellites' (translated by Allen Thomson), https://fas.org/spp/military/program/sigint/androart.htm (accessed 7/05/2019)

Ball, D., *A Suitable Piece of Real Estate: American installations in Australia*, Hale and Iremonger, Sydney, 1980

Ball, D., 'Defence aspects of Australia's space activities', *Canberra Papers on Strategy and Defence*, no. 92, ANU, Canberra, 1992

Ball, D., *Pine Gap: Australia and the US geostationary signals intelligence program*, Allen & Unwin, Sydney, 1988

Ball, D., 'The intelligence war in the Gulf', *Canberra Papers on Strategy and Defence*, no. 78, ANU, Canberra, 1991

Ball, D. and Carr, A. (eds), *A National Asset: 50 years of the Strategic and Defence Studies Centre*, ANU Press, Canberra, 2016

Ball, D., Robinson, W. and Tanter, R., 'Australia's participation in the Pine Gap enterprise', *NAPSNet Special Reports*, https://nautilus.org/napsnet/napsnet-special-reports/australias-participation-in-the-pine-gap-enterprise/ (accessed 7/05/2019)

Ball, D., Robinson, W. and Tanter, R., 'The militarisation of Pine Gap: Organisations and personnel', *NAPSNet Special Reports*, https://nautilus.org/napsnet/napsnet-special-reports/the-militarisation-of-pine-gap-organisations-and-personnel/ (accessed 7/05/2019)

Ball, D., Robinson, W., Tanter, R. and Dorling, P., 'The corporatisation of Pine Gap', *NAPSNet Special Reports*, https://nautilus.org/napsnet/napsnet-special-reports/the-corporatisation-of-pine-gap/ (accessed 7/05/2019)

Bamford, J., *Body of Secrets: Anatomy of the ultra-secret National Security Agency from the Cold War through the dawn of a new century*, Anchor, New York, 2002

Bamford, J., 'The most wanted man in the world', https://www.wired.com/2014/08/edward-snowden/ (accessed 9/05/2019)

Bamford, J., *The Puzzle Palace: Inside the National Security Agency, America's most secret intelligence organization*, Penguin, Harmondsworth, 1983

Barritt-Eyles, L., 'Remembering the Gulf War', *Australian Outlook*, 2 August 2018, https://www.internationalaffairs.org.au/australianoutlook/remembering-the-gulf-war/ (accessed 9/05/2019)

Berkowitz, Dr B., 'The National Reconnaissance Office at 50 years: A brief history', National Reconnaissance Office, Virginia, 2018

Bloggs, C. (ed.), *Masters of War: Militarism and blowback in the era of American empire*, Routledge, New York, 2003

Campbell, D., 'Inside Echelon', *Heise Online*, www.heise.de/tp/deutsch/special/ech/6928/1.html (accessed 8/05/2019)

Central Intelligence Agency, 'The history of SIGINT in the Central Intelligence Agency, 1947–70', https://nsarchive2.gwu.edu/NSAEBB/NSAEBB506/docs/ciasignals_16.pdf (accessed 8/05/2019)

Charlston, J.A., 'What we officially know: Fifteen years of satellite declassification', *Quest: The history of spaceflight quarterly*, vol. 17, no. 3, 2010, pp. 7–19

Cooksey, R., 'Pine Gap', *The Australian Quarterly*, vol. 40, no. 4 (Dec. 1968), pp. 12–20

Curran, J., 'Gough Whitlam's Pine Gap problem', *The Australian*, 5 November 2014

Curran, J., *Unholy Fury: Whitlam and Nixon at war*, Melbourne University Press, Carlton, 2015

Dibb, P., *Inside the Wilderness of Mirrors: Australia and the threat from the Soviet Union in the Cold War and Russia today*, Melbourne University Press, Carlton, 2018

Dibb, P., *The Nuclear War Scare of 1983: How serious was it?*, Australian Strategic Policy Unit, October 2013

European Parliament, Report on the existence of a global system for the interception of private and commercial communications (ECHELON interception system), 11 July 2001, http://www.europarl.europa.eu/sides/getDoc.do?pubRef=-//EP//NONSGML+REPORT+A5-2001-264+0+DOC+PDF+V0//EN&language=EN (accessed 9/05/2019)

Fakley, D., 'The British Mission', *Los Alamos Science*, Winter/Spring 1983, pp. 186–9

Gerecht, R.M., 'The counterterrorism myth', *Atlantic Monthly*, www.theatlantic.com/magazine/archive/2001/07/the-counterterrorist-myth/302263/, July/August 2001 (accessed 9/05/2019)

Greenwald, G., *No Place to Hide: Edward Snowden, the NSA and the surveillance state*, Hamish Hamilton, London, 2014

Griffiths, B., *The China Breakthrough: Whitlam in the Middle Kingdom 1971*, Monash University Publishing, Clayton, 2012

Hager, N., *Secret Power: New Zealand's role in the international spy network*, Craig Potton, Nelson, 1996

Haines, G., 'CIA's role in the study of UFOs, 1947–90: A die-hard issue', https://www.cia.gov/library/center-for-the-study-of-intelligence/csi-publications/csi-studies/studies/97unclass/ufo.html (accessed 9/05/2019)

Harding, L., *The Snowden Files: The inside story of the world's most wanted man*, Guardian Books, London, 2014

Hymans, J., 'Isotopes and identity: Australia and the nuclear weapons option, 1949–1999', *The Nonproliferation Review*, Spring 2000, pp. 1–23

Joint Committee on Foreign Affairs and Defence, *Threats to Australia's Security: Their nature and probability*, Australian Government Publishing Service, Canberra, 1981

Kealy, L., *The Pine Gap Saga: Personal experience working with the American CIA in Australia*, WordPress, 2008

Keefe, P.R., *Chatter: Dispatches from the Secret World of Global Eavesdropping*, Random House, New York, 2005

Kelly, P., *The Unmaking of Gough*, Allen & Unwin, Sydney, 1976

Lindsay, J. and O'Hanlon, M., *Defending America: The case for limited national missile defense*, Brookings Institution Press, Washington DC, 2001

Lindsey, R., *The Falcon and the Snowman*, Penguin, Harmondsworth, 1981

McDonald, H., 'What really happens at Pine Gap', *The Saturday Paper*, no. 128, 1–7 October 2016

National Security Agency, 'The Potentialities of COMINT for Strategic Warning (aka The Robertson Report)', 20 October 1953, https://www.nsa.gov/Portals/70/documents/news-features/declassified-documents/friedman-documents/reports-research/FOLDER_139/41712139075147.pdf

National Security Council, Review of U.S. Policy toward Australia, With NSSM 204 attached, 8 May 1974, https://www.fordlibrarymuseum.gov/library/document/0398/1981992.pdf (accessed 9/05/2019)

Office of National Assessments, 'A preliminary appraisal of the effects on Australia of a nuclear war', https://recordsearch.naa.gov.au/SearchNRetrieve/Interface/ViewImage.aspx?B=7584267 (accessed 9/05/2019)

O'Neil, A., 'Australia and the "Five Eyes" intelligence network: The perils of an asymmetric alliance', *Australian Journal of International Affairs*, vol. 71, no. 5, pp. 529–43

Pedlow, G.W. and Welzenbach, D.E., 'The CIA and the U-2 program: 1954–1974', CIA, 1998

Reynolds, W., 'Rethinking the joint project: Australia's bid for nuclear weapons 1945–1960', *The Historical Journal*, vol. 4I, September 1998, pp. 853–73

Richelson, J.T., 'The CIA and signals intelligence', National Security Archive Electronic Briefing Book, no. 506

Richelson, J.T., *The Wizards of Langley: Inside the CIA's Directorate of Science and Technology*, Westview Press, Boulder, 2002

Rosenberg, D., *Pine Gap: the inside story of the NSA in Australia*, Hardie Grant, Richmond, 2018

Ruppelt, Edward J., *The Report of Unidentified Flying Objects*, Ace Books, New York, 1956

Tait, P., 'What will happen to Alice if the bomb goes off?', Medical Association for the Prevention of War (NT) and Scientists Against Nuclear Arms (NT), 1985

Talbott, S., *Endgame: The inside story of SALT II*, Harper & Row, New York, 1979

Tanter, R., 'Our Poisoned Heart', *Arena*, https://arena.org.au/our-poisoned-heart-by-richard-tanter/ (accessed 9/05/2019)

Tanter, R., 'Possibilities and effects of a nuclear missile attack on Pine Gap', Australian Defence Facilities Pine Gap, https://nautilus.org/briefing-books/australian-defence-facilities/possibilities-and-effects-of-a-nuclear-missile-attack-on-pine-gap/, (accessed 9/05/2019)

Tanter, R., 'The "Joint Facilities" revisited: Desmond Ball, democratic debate on security, and the human interest', https://nautilus.org/napsnet/napsnet-special-reports/the-joint-facilities-revisited-desmond-ball-democratic-debate-on-security-and-the-human-interest/ (accessed 9/05/2019)

United States Space Command, 'AFSPACECOM Desert Shield/Desert Storm Lessons Learned', https://nsarchive2.gwu.edu/NSAEBB/NSAEBB39/document7.pdf (accessed 9/05/2019)

United States Space Command, 'Operation Desert Shield and Desert Storm assessment, January 1992', https://nsarchive2.gwu.edu/NSAEBB/NSAEBB235/25.pdf (accessed 9/05/2019)

Urban, M., *UK Eyes Alpha: Inside story of British intelligence*, Faber, London, 1997

Walsh, J., 'Surprise down under: The secret history of Australia's nuclear ambitions', *The Nonproliferation Review*, vol. 5, 1997, issue 1, pp. 1–20

Weinberger, S., *The Imagineers of War: The untold story of DARPA, the Pentagon agency that changed the world*, Knopf, New York, 2017

Whitlam, Gough, *The Whitlam Government, 1972–1975*, Viking, Ringwood, Vic., 1985

Woods, R., *Shadow Warrior: William Egan Colby and the CIA*, Basic Books, New York, 2013